Our Great Heritage...
from the beginning

Our Great Heritage...
from the beginning

War and Recovery   1913-1929

HISTORY EDITOR

Richard Skolnik

Yale University, M.A., Ph.D.
Associate Professor of History
City College of the City University of New York

★

CREATIVE DIRECTOR AND DESIGNER

Archie Bennett

★

CONTRIBUTING EDITOR

Jesse Stiller

★

EDITORIAL COORDINATOR

Joanne Cozzi Delaney

★

ILLUSTRATOR

Roland DesCombes

Consolidated Book Publishers

NEW YORK ● CHICAGO

Contents

Note to Parents

The two-hundredth anniversary of American independence provides us with a most appropriate moment to look back over the years and examine the key events and issues, as well as notable personalities, that contributed to the distinctive American experience. It is our belief that it is particularly important today, when large numbers of the populace appear to be uneasy and uncertain about what lies ahead, to look back at the American past, not so much for nostalgic purposes, but in order to restore a sense of direction for Americans. This is not to say that we intend to glorify everything that has happened in the United States or that we can uncover clear lessons from the past to use as guides today. To do so would be to distort the facts and would clearly be a disservice to our readers, both youngsters and parents. What we will do is provide examples of the broad range and scope of the American experience so that we today can better understand and clearly visualize the strengths and resources we can draw upon as well as the limitations and the weaknesses that have kept us from fulfilling all of our purposes and achieving all of our objectives.

Why should there be yet another history of the United States? Aren't there enough books on the subject? Don't we already know what happened back then? These are fair questions yet the answer to all of them is no. There will never be a time to stop writing additional books on American history. Strange as it may sound, the past never stands still to those who examine it in the present. This is because new evidence is always being uncovered, and many times these findings force us to change our ideas about past events, issues, and men. Then, too, we are always discovering new ways to use old evidence. Finally, as our own current interests and concerns change, we tend to look at the past in different ways. For example, as the women of today have begun to ask for fair and equal treatment, they have turned to the past and forced historians to examine more closely how women were treated in days gone by and to emphasize the contributions they made in America.

Each of the twelve volumes in this series represents, we believe, a distinct contribution to the history of the United States. We have not attempted to cover all events, all issues, and personalities. Rather we have selected some of the major trends, along with some of the lesser known events and people that have populated the historical stage in America. We have set out not simply to provide a history of policies, elections, and wars—which we believe are too often the primary subject of historical publications—but to look into many aspects of American life. The arts, industry, folklore, religion, diplomacy, race relations—all can help us better explain the complexity and the breadth, and often the contradictions, of the American past. We have chosen not to present the history of America as one great celebration and triumph but to offer praise and self-congratulations when they were due and to criticize and find fault when Americans behaved in a less than praiseworthy fashion. Finally, in order to make these volumes as up-to-date as possible, we have made every effort to introduce some of the more recent findings and theories of historical researchers. Too often books which appear to be current repeat information that has become outdated and no longer represents the considered opinion of those qualified to make such judgments. Because history is changing, we have tried to change with it—without, of course, discarding those long-established truths about the American past.

The art work in these volumes is a unique achievement in itself. We have chosen not to reprint existing pictures or photos, many of which have long served as the "standard" illustrations. Rather we have gathered together many of today's outstanding American illustrators and have asked them for a fresh visual approach to the past. In all instances they have carefully reviewed the incidents under consideration and have taken great pains to reproduce the events visually, with careful attention to detail. We believe that the result is an effective collection of original art that heightens the impact of this lively reexamination of the American past and makes a fitting contribution to the American bicentennial.

Our Great Heritage...
from the beginning

Woodrow Wilson

A University President in the

Woodrow Wilson was an academic scholar turned politician. With a Ph.D. in history and political science from Johns Hopkins and the notable study *Congressional Government* to his credit, he seemed to have a bright academic career. After teaching at Bryn Mawr and Wesleyan University, Wilson was appointed professor at Princeton University, and in 1902 the trustees named him president of the university. He was the first nonclergyman ever to hold that position.

In eight years at Princeton, Wilson won high praise for his educational reforms. He recruited one of the country's most distinguished faculties, revised the curriculum, and instituted a system of guided study and reading for undergraduates. Yet Wilson could not go as far as he wished. He ran into stiff opposition from influential alumni when he proposed to eliminate the eating clubs (exclusive, fraternitylike organizations) on the grounds that they were undemocratic. Wilson fought tirelessly, though unsuccessfully, and made many enemies on campus; the public, however, applauded his efforts. In the end, however, Wilson decided to leave the antagonisms of Princeton for the world of politics.

The New Jersey Democratic machine, looking to head off a revolt by reformers within the party, gave him the nomination for governor. Only mildly progressive, he seemed to be a perfectly safe candidate. But his on-the-stump contact with peoples' problems led him to change his views. Cutting himself loose from the machine that nominated him, Wilson waged a brilliant campaign on a reform program. He won the governorship handily and immediately proceeded to push through the state legislature a series of notable progressive measures. He got a direct primary system, corrupt practices legislation, workman's compensation laws, and state control over the railroads and public utilities.

These successes made Wilson a national figure and a contender for the 1912 Democratic presidential nomination. He further enhanced his reputation during a national speaking tour. Listeners were impressed by this articulate, learned, forceful, transplanted south-

White House

erner. At the convention in Baltimore, Wilson finally won the nomination on the forty-sixth ballot after a good deal of vicious political infighting. With the nomination came almost an assurance of victory, since the Republicans were split between the Roosevelt and Taft factions. Wilson started the campaign slowly but, egged on by Progressive supporters like Louis D. Brandeis, gradually became more outspoken in his calls for reform. He pledged to use federal power to sweep away special privileges and artificial restraints and to restore business competition. This program came to be known as the New Freedom. On election day, Wilson emerged with a distinct plurality (though not a majority) of the popular vote and won by a large margin in the electoral college.

For the new president, implementing his New Freedom policies was the top priority. He immediately sought legislation to revamp the nation's banking system which, under private control, had proven unequal to the task of supplying the currency needs of new business.

Assurance

Certainty; confidence.

Significant

Important; full of
meaning.

Wilson and Representative Carter Glass of Virginia introduced the
Federal Reserve Act to set up a central banking system under the
direction of a Federal Reserve Board. This significant piece of legislation passed Congress in December 1913.

At the same time, Wilson was working hard to gain a reduction in
tariff schedules, a goal that had been blocked for decades. Under the
Republicans, business had been granted protection against foreign
competition in the form of high tariffs. Wilson, believing this to be an
unfortunate and unfair intrusion by government into the free enterprise system, was able to push through Congress the Underwood
Tariff Act, which lowered duties on iron, sugar, wool, and many other
items. He then attempted to regulate business monopolies. The result of this effort was the Clayton Anti-Trust Act and the establishment of the Federal Trade Commission.

Ignore

Pay no attention to.

By the end of his first term, Wilson had accomplished nearly everything he had set out to do. But his legislative programs had practically ignored social problems. Facing a Progressive-Republican
challenge in 1916, Wilson swung from his belief that such matters
were beyond the proper scope of the federal government to a more
affirmative position. On January 18, 1916, Wilson appointed his Progressive mentor, Louis D. Brandeis, to the Supreme Court. With
legislative sanction, the president achieved measures providing
long-term credit to farmers, federal aid for the building and maintenance of highways and schools, and regulation of child labor. So
when Wilson began campaigning in 1916 against his Republican
opponent, Charles Evans Hughes, he was in the enviable position of
having accomplished more than he had promised in 1912. Still the
race was close. Wilson won, again without a majority, largely on the
strength of one issue: he had kept the country out of the European
war.

Yet Wilson knew that the United States might be sucked into the
conflagration at any moment. In February 1917 Germany resumed
unrestricted submarine warfare against all merchant shipping, including the vessels of neutrals such as the United States. When the
Kaiser's U-boats sank five American merchant ships, in March, the
country could no longer remain uncommitted. On April 2, 1917, an
ashen-faced Wilson went before Congress, asking it to declare war
so that the world might be "made safe for Democracy."

Wilson's second term was almost entirely devoted to the war and
concluding a peace. He proved a resourceful war leader and a voice
of idealism in restoring the war-torn world. Wilson's plan for peace,
embodied in the Fourteen Points and the League of Nations, was
ultimately rejected by the American public. As always, he took this
defeat personally and left office in 1920 a physically broken and
bitterly demoralized man.

Triangle Fire

Sweatshop Abuses and Their

Insatiable

Very greedy; that which cannot be satisfied.

A great wave of immigrants—nearly fifteen million in all—came to the United States between 1890 and 1920. Although all ethnic groups were represented, a large percentage of those making the long transatlantic voyages were Eastern European Jews. Set adrift in a strange and frightening land, most of these new Americans gravitated to New York's Lower East Side where they took comfort in the company of their countrymen. The men could usually find only low-paying jobs, and their wives and daughters were often obliged to work. Most unskilled jobs were in the city's bustling garment trade, with its almost insatiable appetite for cheap labor. Workers were paid as little as six dollars for a six-day, fifty-hour week. The owners of these sweatshops often forced their employees to work overtime at night or on Sunday without pay. It was a common practice for employers to charge the operators high rates for electricity and use of the machines.

Powerless as individuals, garment workers came to realize that their only strength was in unity. In September 1909 two hundred employees of the Triangle Shirtwaist Company walked off their jobs in an effort to force management to recognize the new International Ladies' Garment Workers Union (ILGWU) as their sole bargaining agent. Because of stubborn resistance by the factory owners, who hired scab labor and bands of thugs to terrorize the young women walking the picket lines, sympathy for the strikers grew and it quickly spread to the rest of the industry. By the end of November nearly twenty thousand workers had expressed their solidarity with the Triangle strikers by leaving their machines. As the tide of public opinion turned against the owners, they were forced to bargain with the union. By the end of February 1910 the strike was settled. The union had won a victory—working hours were cut and salaries raised—but it had not won recognition for itself or for the principle of a union shop, which would have barred employment of nonunion workers. And some firms ignored their agreement with the union.

At the Triangle Company, where the struggle began, hazardous

Solidarity

Fellowship or union arising from common interests and responsibilities.

Fatal Results

conditions remained uncorrected. And about 4:30 P.M. on Saturday, March 25, a fire broke out. Nearly five hundred workers were trapped on the eighth, ninth, and tenth floors. Panic-stricken, they stampeded to the exit doors, but all but one of them had been bolted shut, a measure employed by the owners to prevent pilferage by the workers. Many workers packed themselves into the one working freight elevator; others threw themselves down the shaft at the plummeting car. Firemen began the grim task of removing the charred remains of the one hundred workers who were not able to get off the work floor. All but twenty-one of the victims were women, most of them Jewish.

A mass funeral ceremony on a rainy April 5 was attended by over one hundred thousand people, expressing their grief and demanding that the government take steps to prevent another disaster from occurring. An investigation by a state commission led to the enactment of fifty-six laws designed to protect workers. The deaths had not been entirely in vain.

Armory Art Show

The "Lunatic Fringe" in American

Deride

Laugh at in scorn; make fun of.

While the American gift for invention was admired the world over, the Europeans derided the relative lack of American culture. Nowhere was the cultural vacuum so glaring as in the field of painting. After a century and a half of political independence, Americans had still not liberated themselves from the self-deprecating opinion that the only true art was European art. Young American painters went to Paris and London to study at academies that had instructed countless generations of painters in the traditional manner. However much it stifled spontaneity and creativity, formalized study abroad at least gave the struggling artists the necessary credentials to procure the financial support of a sponsor. With the patronage of newly rich Americans, artists turned out innumerable canvases patterned on the Old Masters, works of little merit that conformed to the accepted aesthetic theory. Artists who refused to reproduce sterile forms found themselves on the streets, trying to sell their canvases to a largely oblivious public.

Robert Henri was determined to end this sad state of affairs in American Art. A native of Cincinnati who had studied in Paris, Henri became disgusted with an art establishment that compelled him and his many able colleagues to divert their talents from "expressing the spirit of the people" to more or less copying the style of the Old Masters. When the jury of the respectable National Academy of Design in New York refused to consider the paintings of Henri and seven of his friends in its competition, they opened their own exhibition of "The Eight" in February 1908. Visitors found none of the stiff portraits or rustic landscapes to which they had become accustomed. They

Depictions

Portrayals; paintings.

saw instead vivid, sometimes grim, depictions of American life that led scandalized critics to dub these painters the "Ashcan School."

While the art world was still buzzing from "The Eight" exhibition, a number of enthusiasts undertook to bring the new realism in art to a wider audience. They planned a show on so grand a scale that it had to be held in New York's vast Sixty-ninth Regiment Armory.

The Armory Show opened on February 7, 1913. Nearly sixteen

Painting

hundred paintings, the work of over three hundred artists, were displayed. Many of these works were unorthodox in conception or execution. Among them were the canvases of the "heretical" French Impressionists, who were then causing a sensation in Paris. The critics, among them Theodore Roosevelt, had a field day; the former president coined the term "lunatic fringe" to describe the new wave of painters. Despite the outpourings of shock and indignation from the art establishment, thousands flocked to the armory and patiently stood in line to see Marcel Duchamp's *Nude Descending a Staircase* or one of the other controversial works. From New York the exhibition traveled to Chicago and Boston, creating a stir in both cities. In Chicago, a self-appointed guardian of the public morals, outraged by the sarong-clad Polynesian girls prominently displayed in Henri Matisse's paintings, called out the Vice Commission to investigate the show. Matisse was later burned in effigy by some of the city's more conservative art students.

Indignation

Anger at something mean, unfair, or unworthy.

17

Red Grange

"The Galloping Ghost of the Gridiron"

Gridiron

A football field.

Boisterous

Rough; violent.

The twenties had more than its share of legendary sports figures. There was Babe Ruth, the Sultan of Swat; Jack Dempsey, the Manassas Mauler; and Paavo Nurmi, the fleet-footed Finn. Each of them was instrumental in restoring a lost image, or creating a new one, for his sport. They each possessed a certain quality of character, an appeal which high-spirited Americans of the twenties found irresistible. Their exploits, on and off the field of competition, heroic and all too human, filled countless newspaper columns, lured throngs of spectators, and left fond memories for every fan privileged to watch them. Harold "Red" Grange, the "Galloping Ghost" of the gridiron, did all of this and more. Like Ruth and Dempsey, Grange was almost single-handedly responsible for turning a game of marginal interest into a viable professional sport. But unlike Ruth and Dempsey, whose manner some found offensively boisterous and whose habits, as they themselves admitted, were scarcely worthy of emulation by the young, Grange seemed to embody all the old virtues. He was modest and reserved, frugal and ambitious, clean-living, and above all one of the most gifted athletes ever. And while Ruth cut a slightly ridiculous figure on the field, paunch hanging over his belt, and Dempsey was basically a brawler who, had fame in the ring eluded him, would have been perfectly at home in back alleys, Grange was every inch the athlete. In high school he excelled at basketball, baseball, football and track.

Arriving on the University of Illinois campus in 1922, unsure of his abilities and overawed by the presence of Coach Bob Zuppke and his collection of stars, Grange toyed with the idea of not even trying out for the team. A 65-yard run through the entire varsity on his first practice play dispelled all his doubts and convinced Coach Zuppke that Grange was a man with unique talents. In his first varsity game he ran for three touchdowns, and accumulated nine more before the season was out. Grange was virtually unstoppable. He had a knack for finding holes where none seemed to exist, and for foiling would-be tacklers with impossible moves and sudden bursts of speed.

18

Yet, even as he piled up records and ecstatic press clippings, his renown was limited to the midwest. Eastern football fans remained skeptical of this much-touted collegian. After all, he had never proven himself against the powerhouse teams of the Ivy League. So when the 1925 schedule matched Illinois against mighty Pennsylvania, the stage was set for a contest that would either vindicate or destroy Grange's reputation. It took place on a rainy afternoon in October, at Philadelphia's Franklin Field before a capacity crowd of sixty thousand. Grange proved he was no fluke. Taking a hand-off on the initial play from scrimmage, he proceeded to scamper fifty-five yards for a touchdown. He scored twice more on that gloomy day, rushing 363 yards on a muddy field in leading underdog Illinois to a 24-2 victory.

By the time of Grange's last appearance in an Illinois uniform, he had already signed lucrative contracts with a personal manager, C.C. "Cash and Carry" Pyle, and with George Halas, owner of the Chicago Bears of the infant National Football League. Many fans believed that Grange had made a fatal mistake, for in 1925 "professional" football labored in near obscurity. It was a dismal affair. Ragtag groups of semiskilled men played—often on two different teams—for fifty or seventy-five dollars a game at places like Pottstown, Pennsylvania, Canton, Ohio, and Rock Island, Illinois. Schedules were rarely adhered to; more often, rival teams arranged a game whenever it was mutually convenient. But when Red Grange entered the league he brought with him his legion of admirers, his impeccable reputation, and his unmatched skills. On the strength of his name, the Bears drew overflow crowds wherever they played, and the enthusiasm spread to other cities as well. In his first year of pro ball, the Galloping Ghost earned close to half a million dollars, largely from a unique salary arrangement that gave him a percentage of the box office receipts. He endorsed a variety of products and contracted to make a movie. Grange brought football into the modern age.

19

The First Tarzan

Burroughs Captures the

Swinging from vine to vine, in the thicket of an African rain forest, is seen the well-muscled body of a human, clad only in a leopard-skin loincloth. Suddenly he gives forth a blood-curdling howl, and the animals of the jungle, rather than scurrying as they would if faced with danger, merely pause in their movement. For this is Tarzan, their friend and defender, born of man, but raised by apes.

Since 1912 that scenario has captivated the minds and hearts of people the world over. The adventures of Tarzan possess universal appeal, incorporating the elements of suspense, intrigue, and the unknown. He has long been among the most durable characters of fiction. For years all that people knew (or thought they knew) of Africa they learned from reading about Tarzan. Ironically Tarzan was the invention of a man who had never ventured even close to the so-called Dark Continent, a man whose familiarity with his subject was spotty at best and largely derived from other fictional works and a fertile imagination.

The author was Edgar Rice Bur-

Book

World's Imagination

roughs. Burroughs, a one time cowboy, soldier, policeman, goldminer, storekeeper, aluminum salesman, and advertising man became convinced that he could write popular fiction. He turned first to science fiction. The result of his labors was "Under the Moons of Mars," which was serialized in the magazine *All Story* in 1912. This was followed by another Martian tale, which also received a warm reception. Encouraged by his success, Burroughs then turned to a topic that had long held a fascination for many people. One sleepless night, Burroughs came up with the idea of a child, abandoned in the jungle, adopted and reared by apes. He later stated that he conceived of this tale as a variation on the Romulus and Remus legend. It is also probable that he was influenced by the works of Rudyard Kipling. He completed *Tarzan of the Apes* before the year was out. The response was so favorable that the A. C. McClurg Company put it out in book form, and everyone clamored for more.

In spite of its success, Tarzan never became a fixation for Burroughs. The breadth of his interests was as remarkable as his enormous creative output. Among his fifty-nine books were *The Cave Girl*, an imitation of Jack London; a Hollywood exposé, *The Girl From Hollywood*; a business novel, *The Efficiency Expert*; and a number of imaginative works of science fiction.

But the Tarzan stories were always his bread and butter. He wrote twenty-four of them, which sold in excess of forty million copies. His characters were unforgettable: Tantor, the elephant; Numa, the lion; Hista, the great snake; and Kala, Tarzan's ape-mother. He managed to turn his virtual ignorance about things African into an asset. By creating his own world, with its own history, language, psychology, and genetics, Burroughs escaped the pitfalls of having to confine and order his characters around facts. For Burroughs to have tried to rely on actual details would have been to deprive the stories of their vitality, changing them, as one historian has pointed out, from among "the most imaginatively real creations of their kind" into "only poor travelogues."

Breadth

How broad and far a thing is.

Figment

A fiction; something imagined.

Panama Canal

The Dawning of a New Age

On the morning of August 15, 1914, the American freighter the S.S. *Ancon* lay just off the city of Colón on the Atlantic side of the Isthmus of Panama. At a few minutes past eight the ship hoisted anchor and set sail on a brief but historic journey. Although the *Ancon* had been prettied up a bit, there were no speeches or extravagant displays. Yet this short voyage (it took only seven and a half hours) marked the culmination of one of man's greatest and, for forty years, most elusive undertakings: joining the waters of the two oceans via the Panama Canal. The *Ancon* was the first ship to pass through this waterway.

Hoist
Lift up.

The construction of the Panama Canal had not been easy. A French company tried and failed when it ran into money problems and the malarial jungles of Panama. Then the isthmus was wracked by years of revolution and chaos. But President Theodore Roosevelt was preoccupied with the idea of progress and decided that the world could no longer wait until the Panamanians settled their domestic troubles. According to Roosevelt's plan, the United States would do the job. Roosevelt engineered a neat coup d'état to gain an American protectorate over the ten-mile-wide Canal Zone. Then he set up a seven-member Panama Canal Commission to study all aspects of the enterprise. Finally work began.

Coup d'etat

A decisive and sudden measure in politics.

The first job was to wipe out the yellow fever and malaria that had taken such a toll of the French during the 1880s. Roosevelt appointed an expert on military sanitation, William C. Gorgas, as the project's medical officer. Gorgas knew that these two dreaded diseases, which could destroy a work force in a matter of days, were carried by mosquitoes that bred in swamp lands. Gorgas soon had men all over the Canal Zone digging drainage ditches and spraying oil over stagnant water to suffocate the mosquito larvae. After digging over seventeen hundred miles of ditches, spraying a half million gallons of oil (larvacide), and distributing thousands of quinine pills (an effective antimalaria drug), Gorgas had the problem well under control.

Until 1906 there was little work done on the canal itself. Congress had not yet decided which of several construction plans to adopt. In

Completed

of Shipping

the meantime John F. Stevens, the chief engineer, had begun building up the Panamanian railroad. Congress finally decided to authorize a canal based on a series of locks. (A lock is an enclosed chamber that either raises or lowers ships by regulating the water level; they were needed because inland the isthmus was several hundred feet above sea level.) Work then proceeded under a new chief engineer, George Washington Goethals.

It was up to Goethals, a quiet, intense colonel in the Army Corps of Engineers, to coordinate every aspect of the huge project. It required prodigious energies and determination. For one thing, there was the question of labor. Lured by wages ten to twenty percent higher than were being paid in the United States for comparable work, nearly forty thousand laborers poured into the Panamanian camps. Goethals ran the Canal Zone with military efficiency, molding this force into a disciplined body. He would not allow any opposition to stand in his way.

There was a great deal of grumbling among the men about Goethals's stern policy, but considering the enormity of the mission, he had little choice. Although the technology to build this type of canal was not new, no one had ever attempted to perform such an engineering feat. Each section of the canal posed its own problems. The Culebra Cut, at the Atlantic end, involved slicing through nine miles of mountains. Almost twenty-one million cubic yards of dirt and rock were eventually removed. Gatun Dam, a mile and a half long and almost half a mile thick at the base, enclosed Gatun Lake, the biggest man-made body of water in the world (164 square miles). Six pairs of locks, each 1,000 feet long, 110 feet wide, and 40 feet deep, were built using 4.5 million cubic yards of concrete. The gates, made of steel, were seven feet thick. Nature at times intervened in the form of landslides, burying weeks of work in a few seconds. But after eight long years of construction, and a cost to the government of more than $375 billion (it has been estimated that the same project would cost $3 billion today), the canal was ready for the *Ancon*.

Isthmus

Long, narrow piece of land with water on both sides.

Opposition

Resistance; action against.

23

Ford Assembly Line

Mass Production of Automobiles

As the twentieth century dawned, almost all Americans still traveled on horseback. Gasoline-powered buggies had been on the market for several years, but sales were sluggish. The first cars were noisy, unreliable, and expensive (they cost over two thousand dollars). And in 1900 the car was neither a practical nor pleasurable mode of transport. There was not a single service or filling station in the entire country, and paved roads were almost nonexistent except in the East. Many a leisurely Sunday outing turned into a grimy struggle to free a car stuck in axle-deep mud.

But Henry Ford was determined to manufacture a car that every family could afford—and, as important, a car that was simple to maintain, cheap to operate, and built to last. The trick, as Ford saw it, was to streamline the assembly process and thus turn out thousands of cars that were indentical. The technology already existed. Ransom E. Olds (creator of the Oldsmobile) had developed techniques in mass-producing autos; and Henry Leland had demonstrated that cars could be made with interchangeable parts. Ford combined both principles and, with a few modifications and a number of innovations, applied them in constructing an efficient and profitable plant.

Integral to Ford's plan was the assembly line—a conveyer belt that

Sluggish

Not active; slow moving.

Becomes a Reality

moved at a constant rate before lines of workers, bringing their work to them so that not a single movement was wasted. Ideally, a piece of iron would never stop moving from the time it was forged until it became a part of the finished product. Before this a single skilled worker had constructed each part of the automobile. He then brought the finished part to a central area where it was attached by hand to the car's skeleton. For example, assembling a magneto (a device that controls the ignition) called for painstaking precision and generally took about twenty minutes. On the Ford assembly line, however, each worker performed only one task—adding or soldering a wire, for example. Assembly time for the magneto was soon reduced to five minutes. The entire vehicle was created on the assembly line, and the 1908 Model T sold for the unheard-of price of $850. By the mid-1920s, because of increased efficiency on the assembly line, Ford was able to cut the price to less than $300. By May 1927 fifteen million Model T's had been produced. No options were available on the Model T; Ford once said that "any customer can have a car painted any color he wants so long as it is black." Ford succeeded in democratizing the automobile, and in doing so, he changed America into a mobile society.

Precision

Being exact; accuracy.

Federal Trade

First Steps to Regulate Big Business

Advocate

Recommend publicly; speak in favor of.

To regulate or not to regulate big business, that was the question—and if so, how? Over no other issue did emotions run so high nor was opinion so polarized in the decades after 1870. Depending on whom one was listening to, business unchecked would either lead to oligarchy and the ruin of democracy or unsurpassed national grandeur. While the militant antibusiness forces advocated the nationalization of key industries, their opponents urged the government to keep its hands off the free enterprise system, except to promote the growth of American industry (through protective tariffs, for example). What emerged from over three decades of heated debate was a compromise—the Federal Trade Commission Act of 1914 which created an agency to act as a watchdog over business, to root out its most glaring excesses and abuses and thereby to restore a good name to the vast majority of businesses that deserved it. This satisfied all but the most vocal critics of capitalism, for most Americans believed that the system needed only to be cleansed of a few objectionable aspects for it to operate in the country's best interest.

Ultimate

Final; last.

Americans believed there was good reason to exercise a tighter rein over big business. One of the most serious things it did was to stifle and ultimately eliminate most smaller competitors. By 1900 the independent small producer was a thing of the past in many basic industries. For example, giant corporations had risen to control virtually the entire production of such items as steel, oil, sugar, and tobacco. These were probably the most efficient and profitable enterprises ever known, and they contributed immeasurably to the country's high standard of living and its status as a world power. But such efficiency came at high cost. It meant that countless small businessmen were wiped out and that the prices of many monopolized consumer goods and services—such as those supplied by the railroads and the banks—were set from above. It meant corruption in government, as the captains of industry spread money around liberally to gain favorable treatment from legislators. It meant that in the end control of the economy rested not with the people but with the

Commission

financiers and the corporations.

The first public stirrings against monopoly occurred in the 1870s among farmers who cried out for relief from discriminatory railroad freight rates and high interest charges on loans. Organized into the Populist Party, whose demands included public ownership of the railroads, the farmers won several western and southern states over to antitrust legislation. But the railroads and the industrial monopolies were essentially national institutions and, as such, only the federal government could adequately regulate them. With public agitation mounting, Congress passed a host of measures designed to place the railroads under control (for example, Interstate Commerce Act of 1887) and to abolish rate schedules which discriminated against the small shipper (Hepburn Act of 1906). In 1890 the government passed the Sherman Anti-Trust Act as a weapon against monopolistic restraint of trade or commerce. Under Presidents Roosevelt and Taft, the Justice Department brought suit against nearly seventy firms which were charged with violations of the Sherman Act. As the result, such corporations as the Standard Oil Company and the American Tobacco Company were forced to give up some of their huge holdings.

Yet many big businesses found ways of getting around the law. In 1912 the government sought to close loopholes in the Sherman Law with another measure, the Clayton Bill, which contained a long list of specific proscribed business practices. But it was impossible to enumerate all the possible questionable practices which business might employ. That was when President Wilson threw his support behind a bill outlawing unfair trade practices in general terms, which created a regulatory commission to oversee the operations of large corporations. Wilson signed the Federal Trade Commission Act in September 1914. Although the commission was subsequently attacked for being too friendly to business, it won a long-standing battle against the Beef Trust and contributed to restoring the reputation of big business in the United States.

Adequate
Enough; sufficient; as much as needed.

Enumerate
Count; find the number of.

The U.S. and Mexico

Relations with an Unstable Republic

Abrupt

Sudden.

In May 1911 the thirty-four-year reign of Mexican dictator Porfirio Díaz was brought to an abrupt end. In a virtually bloodless coup, power passed into the hands of Francisco Madero, a popular reformer who promised to abolish the privileged positions which the Catholic church, the army, the wealthy landowning class, and foreign capital had enjoyed under Díaz. Naturally those interests resisted Madero, and in February 1913 Madero was himself deposed in a conspiracy headed by his army chief, Victoriano Huerta. The new regime was accorded recognition by the European powers; in Washington, however, outgoing President William Taft decided to leave the question of recognition to his successor, Woodrow Wilson.

Wilson, who refused to put the stamp of approval on a regime that had come to power by undemocratic means. The leader of the growing anti-Huerta movement, Venustiano Carranza, more nearly fitted Wilson's idea of the kind of ruler that Mexico needed. When it became clear that Huerta was incapable of suppressing the northern insurgents, Wilson proposed that the United States work out a settlement between the federalist (Huerta) and the constitutionalist (Carranza) factions. This offer was rejected by both sides, who resented American meddling in Mexico's internal affairs.

Emissary

A person sent on a mission or on an errand.

With the constitutionalists gaining ground, Huerta dissolved the National Assembly and arrested most of its members. Wilson, appalled at this "usurpation," now decided that Huerta must be removed. The president sent an emissary to Carranza, promising him American aid in his battle against the federalists. Carranza rejected these proposals, and he and his followers made it clear that they would resist U.S. intervention, military or otherwise.

Here matters stood until April 10, 1914, when a trivial incident provided Wilson with what he thought was the perfect pretext to use military force to unseat Huerta. An overzealous *Huertista* colonel arrested several crewmen of an American whaleboat when they unwittingly landed in a restricted area at Tampico. The men were quickly released and an apology was sent to Admiral Henry T. Mayo.

But this apology was not enough for the admiral, who demanded a twenty-one-gun salute to the American flag. Huerta refused. Wilson accused Huerta of deliberately insulting the United States and ordered marines ashore. By April 22 Veracruz was in American hands. The mission had cost ninety American and three hundred Mexican lives. It accomplished nothing other than to inflame Mexican hatred for the United States. The *Carrancistas* were poised for victory anyway, and on August 20 they drove Huerta from office.

No sooner had Carranza taken office than a new rivalry broke out between him and his best general, Francisco "Pancho" Villa. Wilson threw his support behind Villa, although the general was ignorant and unfit for political leadership. The ensuing struggle saw Villa's armies so badly beaten by the *Carrancista* forces that he had to seek refuge in his native stronghold in Chihuahua Province. Wilson finally extended recognition to Carranza.

Villa was down but not out. On January 11, 1916, a band of *Villistas* stopped a train at Santa Ysabel, and shot sixteen Americans on the spot. In March his men crossed the border and raided the town of Columbus, New Mexico, killing nineteen inhabitants. By these bold actions Villa hoped to draw the United States and Mexico into war, which he could then exploit to his own advantage. It seemed as though his plan would work when Wilson ordered an expeditionary force under General John Pershing to capture the Mexican bandit. The Carranza government began to grow suspicious that the Americans were less interested in apprehending Villa than in occupying Mexican territory. Skirmishes broke out between Mexican and American troops, and Carranza then demanded withdrawal of the expedition. Wilson refused and called up one hundred thousand National Guardsmen. Tempers were frayed on both sides, but neither wanted war. Carranza ordered the release of the American prisoners, and a joint high commission was set up to iron out differences between the two countries. The darkest chapter in Mexican-American relations was drawing to a close.

Rivalry

An effort to obtain something that another person wants.

Apprehend

Seize; arrest.

31

The Sinking of

An End to American Neutrality

Impartial

Showing no more favor to one side than to the other.

Americans were deeply shocked and distressed at the outbreak of war in Europe. President Wilson, however, urged them to be "impartial in thought as well as action" and comforted them with the suggestion that the war was one "with which we have nothing to do, whose causes cannot touch us." Despite the president's lecture, Americans could not remain entirely impartial. Ethnic groups identified with their fatherlands; large numbers felt deep ties with Great Britain, the repository of the Anglo-Saxon heritage. But while most Americans aligned themselves emotionally with one side or the other, the country as a whole was content to be a passive observer of European events.

Belligerent

At war; fighting.

Neutrality of action would soon become as difficult to maintain as that of thought. With the belligerent armies deadlocked in the trenches of France, the focus shifted to the struggle on the high seas. Britain, seeking to starve Germany into submission, had clamped a blockade on shipments of strategic materials to her enemy. In violation of international law, ships carrying contraband were stopped and searched, and their cargoes were seized. American vessels were not immune to such treatment. Wilson and his advisers protested, but they were sympathetic to the Allied cause and were unwilling to deprive Britain of her naval superiority at a time when Germany had the upper hand on the battlefield.

A more ominous threat to U.S. neutrality was the U-boat, widely employed for the first time. Von Tirpitz, Germany's secretary of naval affairs, saw that the submarine, which could be built cheaply and carried a small crew, might offset his country's surface fleet inferiority and prove useful not only in breaking the British blockade but also in raiding Allied shipping. Since Britain was wholly dependent on imports for food and fuel, the U-boat could be a decisive force in the war. The deadly potential of the submarine was demonstrated in September 1914 when the *U-9* sank three British battle cruisers in a single day. Encouraged by this unexpected success, the Germans began a crash submarine-building program and unleashed these

the Lusitania

vessels on commercial shipping in the Atlantic.

International law was quite specific on sea warfare, especially when it concerned neutral merchantmen. Belligerent ships could stop their prey for search, and they could confiscate or destroy cargo destined for an enemy. But they had no right to destroy the ship and were required to provide for the safety of all persons aboard. Adherence to these restrictions would have been fatal for the submarine because above water it was a thin-skinned, nearly defenseless sitting duck. Its value was in the surprise attack it could mount, and since enemy merchant ships regularly flew neutral flags as a disguise, there was no practical way of telling one from the other. Mistakes—fatal mistakes—were inevitable.

On Saturday, May 1, 1915, the majestic British liner *Lusitania* sailed from New York with 1,956 passengers and crewmen aboard. She also carried 4,200 cases of rifle cartridges in her hold for delivery in Liverpool. After seven days at sea, she was sighted off the Irish coast by the German submarine *U-20*. At 3:10 P.M., Lieutenant Commander Schwieger of the *U-20* ordered the delivery of one torpedo. It spun through the water, striking amidships with a tremendous explosion and ripping a huge hole in the *Lusitania's* starboard side. Passengers panicked, jamming lifeboats so full that they sank immediately on hitting the surface. Within fifteen minutes the great ship *Lusitania* began to sink. The tragedy cost the lives of 1,198 people, including 128 Americans.

Americans were horrified by the sinking, yet there was no immediate outcry for war. "There is such a thing as a man being too proud to fight," declared Wilson. Three strong communiqués to Germany resulted in a German pledge to refrain from further attacks on passenger ships, and for the moment the American people were satisfied. But later in the war, with Germany's military situation growing more desperate, she resumed unrestricted attacks on all ships in the war area. With American ships being sunk almost weekly, Wilson had to abandon neutrality.

Confiscate
Take and keep.

Inevitable
That which cannot be avoided; sure to happen.

Zimmermann

Discovery of Germany's War

Diligent

Industrious; not lazy.

In the 1916 presidential election the American people voiced their desire for peace. Woodrow Wilson was reelected on the grounds that he had "kept the country out of the war." Wilson took his peace mandate seriously. Throughout 1916 he worked diligently not only to preserve American neutrality but also to bring the belligerent powers to the conference table. Germany, whose war effort was going well, could afford to be receptive to Wilson's peace overtures, but only as long as her armies were allowed to remain in already-occupied territories, like Belgium. In truth, the Kaiser was interested in a negotiated victory: he wanted a large indemnity from Britain and France and the end of British naval supremacy. To accept such "peace" terms would be tantamount to surrender for the Allies, who had ideas of their own—they sought nothing less than the destruction of German military power. The neutral world finally learned that the war was being fought for nothing more noble than political and economic supremacy and military glory. Since Wilson's efforts as a mediator met only with frustration, all he could do was to take steps to ensure that this country remained neutral.

Mediator

One who comes in to settle a dispute between two sides or people.

Yet events were developing that were destined to doom even that effort. Four months after Wilson requested an expression of aims from the warring powers, the United States was sucked into the fray. With national survival at stake, the belligerents were adopting increasingly extreme measures to bring the enemy to its knees. The British intensified their economic warfare against nations still trading with Germany. On January 31, 1917, the Germans announced they would abandon the Sussex Pledge to spare unarmed, neutral merchantmen from destruction by U-boats and resumed unrestricted submarine warfare. Germany knew this would bring the United States into the war against it but was hopeful that the stepped-up U-boat campaign would bring victory before American military strength could be brought to bear in the conflict. The result of the German declaration was virtually to drive American shipping off the Atlantic. Lest they be exposed to the renewed U-boat peril, passenger

Note

Intentions

cruises were cancelled and merchant ships stayed in their berths. All the while, goods piled up in wharves and warehouses. Wilson severed diplomatic relations with Germany, but only after a long and agonizing contemplation of the horrors of war. Until February 1917 the American people continued to hope that Germany would commit no overt act to force them into war.

On February 25, Wilson received a message from his ambassador in London which put an end to any doubts as to German intentions. More than a month before, the German Foreign Secretary Arthur Zimmermann had sent a dispatch to his minister in Mexico, which was intercepted by the British, proposing an alliance between the two countries should the submarine attacks provoke war between the United States and Germany. For its troubles Mexico was to receive financial assistance and support in reacquiring the "lost territory in Texas, New Mexico, and Arizona." American officials were astounded and angry. Wilson made the note public during the congressional debate on the question of arming American merchant ships. Though the Senate refused to authorize such "provocative" measures, for the first time the American public came to the realization that war was imminent.

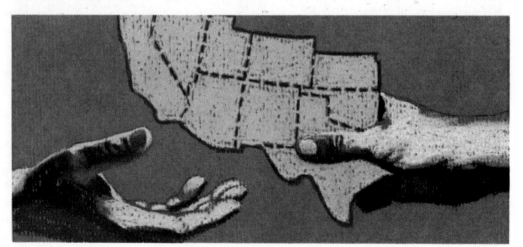

U.S. War-Europe

Making the World Safe for Democracy

Desperate

Having little chance for hope; very dangerous.

Barren

Not producing anything.

When Congress declared war on Germany in April 1917, the Allied military situation was desperate and growing worse. A French offensive had just failed, and ten divisions of her war-weary troops had mutinied. Italy was on the verge of collapse. The Germans were confident that the war would be over before a significant number of American soldiers could be made available to shore up the sagging Allied lines. Her U-boats were exacting a deadly toll on Allied shipping—881,000 tons were sunk in April alone. At that rate, Britain would soon be isolated and starving.

At first, neither Wilson nor his advisers understood the gravity of the situation. They thought that American shipping, credit, and material would be sufficient to bolster the Allied war effort. But when it became clear that the Allies were in very real danger of defeat, the focus shifted to raising, training, and equipping an army large enough to turn the tide against Germany. Conscription was the answer to the first problem. After a bitter debate, Congress approved a selective service act, and by June 17, 1917, nine million men between the ages of twenty-one and thirty-one were registered. Equipping them was another story. Although small arms were relatively plentiful, American arsenals were nearly barren of artillery, tanks, and aircraft. Production was slow in getting underway, and for the most part American forces in Europe were dependent on the French for their heavy arms.

Wilson named Major General John J. Pershing, fresh from the punitive expedition against Villa in Mexico, to head the American Expeditionary Force (AEF) in Europe. Both men realized the urgent need to get even a token force of American troops to Europe as quickly as possible in order to give the beleaguered Allies a psychological lift. To the accolades of the relieved French citizens, Pershing arrived in Paris on June 14, 1917. There were 14,500 American troops in Europe that summer, 300,000 by March 1918, and more than two million by the following November. At the beginning, the Allied field commanders insisted that the American forces be integrated into existing units under their command. With Wilson's assent, Pershing stubbornly held out for keeping the American troops separate.

Thrown into the line at Chateau-Thierry and Belleau Wood, the green American troops performed most admirably. They hurled back the German advance and inspired the British and French to do likewise. The Germans made one all-out attempt to break through to Paris in July, but 85,000 Americans parried the German thrust and began the counterattack that convinced the Germans they were finished. In September, half a million Yanks wiped out a German army at St. Mihiel, taking 16,000 prisoners. In the last major battle of the war, the forty-seven-day Meuse-Argonne offensive, 1,200,000 American troops drove back the Germans.

The war took the lives of 112,000 Americans, with another 2,900 listed as missing. (There were 1,700,000 Russian, 1,385,000 French, and 947,000 British casualties.) Yet it was the American contribution at the decisive moment that defeated Germany and the Central Powers.

U.S. War-America

The American Homefront during

Enormous
Huge.

Americans knew that financing the war would be costly; the question was how the enormous sums would be raised. Conservatives advocated recourse to borrowing and consumption taxes. Progressives insisted that the rich should pay high income, inheritance, and excess profits taxes. Congress chose to walk a middle line. It first authorized a bond issue of $5 billion, though by the end of the war that figure had increased almost fivefold. Over $10 billion was raised by taxes, many of them new. The Revenue Act of 1918 put most of the burden on large incomes and profits. Thus, while real income for the average worker or farmer shot ahead during the war, that of corporations and the wealthy experienced a relative decline.

In order to keep America's industrial plants operating at full efficiency, the War Industries Board was created by Congress. It was supposed to allocate resources, control production, and oversee labor relations. But it became ensnarled in red tape, and Wilson acted decisively. He appointed Wall Street broker Bernard Baruch to head the board and gave him sweeping powers over the entire economy. Baruch juggled scarce resources and set production schedules, martialing American industry to the call of war.

Inaugurate
Begin; start.

When America entered the war, Wilson was informed that food supplies for civilians on the British Isles would last no more than six to eight weeks. On May 19, 1917, the president inaugurated a food control program under the able direction of Herbert Hoover. Hoover's job was to encourage conservation and stimulate production of food. The first end was accomplished with a massive public relations effort and a rationing system. Restaurants served shark steak and whale meat, and mothers were urged not to allow their children to throw their apples away after only a bite or two: "Nowadays even children must be taught to be patriotic to the core." Hoover's tactic in the commodity market was to set such high prices that farmers eagerly increased production. Thus the production of hogs, a staple in the Allied diet, nearly doubled.

Shipping was the Allies' lifeline, without which the food grown on

World War I

American farms could never reach European mouths. American builders had to keep up with the growing demand, made even more acute by the toll taken by Germany's submarines. This area witnessed the greatest failure of the war effort. After much confusion, construction of shipyards did not get underway until the fall of 1917. Thus, the first vessel from the largest government shipyard (in Philadelphia) was not delivered until after the war was over.

Labor thrived during the war years. All the government administrative boards recognized the right of collective bargaining; they compelled adoption of an eight-hour day where it was practical and set wages high enough so that even with the wartime inflation the real income of labor increased about fourteen percent. Unemployment was virtually unknown.

Compel
Bring about by using force.

One of the first of the myriad wartime agencies was the Committee on Public Information, headed by George Creel. The job of the Creel Committee was to counteract the influence of Socialists, certain Progressives, and most Irish- and German-Americans who were at best lukewarm in their enthusiasm for the war. It mobilized 150,000 lecturers and writers in the greatest propaganda campaign in American history. It worked so well, in fact, that Americans became caught up in a war hysteria. Everything German was shunned: sauerkraut became "liberty cabbage." Many states forbade the teaching of the German language. In the Midwest, vigilante groups conducted reigns of terror against German-Americans. In addition President Wilson asked Congress to pass the Espionage Act of 1917. It provided stiff fines and jail sentences for a wide variety of offenses against the war effort. Primarily it was used to stamp out radical criticism. For example, Socialist leader Eugene V. Debs was sentenced to ten years in prison for expressing his revulsion at the war. The law allowed the Postmaster General to deny the use of the mails to any "treasonous" matter. As one historian has written, "In retrospect, the war hysteria seems the most fearful price that the American people paid for participation in the First World War."

Myriad
A great number.

Federal Income

An Answer to the Problem of

We tend to think of the income tax as an inescapable fact of life that has always been with us and always will be. While we can be reasonably certain that the income tax is here to stay, it has not always been so and, in fact, only recent generations of Americans have been subject to it.

The Founding Fathers, had opposed the imposition of a progressive tax (that is, one whose rate increases in proportion to income). The tax burden should to be shared equally and should not relate to the ability to pay. What concerned them was the disturbing possibility that, acting through their representative assemblies, the unruly mass of ungrateful citizens might throw deference aside and attempt to "democratize" the nation's wealth by passing levies that discriminated against their economic superiors. In fact, the Constitution forbade the institution of a graduated income by stating that "no direct tax [may] be laid unless in proportion to the census."

A century later the same concern was in the minds of the wealthy. During the Civil War they had grudgingly gone along with modest tax on incomes, but only under the condition that it would lapse after the emergency was over. After that, the privileged would hear no more of such radical departures. The income tax was roundly denounced in the conservative press as "odious, vexatious, inquisitorial, unequal," and incompatible with a free people.

Meanwhile, the government had reverted to its customary practice of raising revenue almost entirely from indirect consumption, or "use" taxes—customs duties and excise taxes on tobacco and alcohol—which hit the middle and lower classes of consumers hardest. With no federal tax on income or property, enormous individual fortunes were made during the "Gilded Age." The upper crust solidified its hold on the country's wealth; each year fewer of the nation's people controlled ever larger shares of its income. Severe depressions in the 1870s and 1890s brought unemployment and misery to workers and farmers, but not to the Park Avenue patricians, who flaunted their wealth even more openly than ever.

Tax

Revenue Raising

Gradually the idea of an income tax, with its mild leveling effect on society, became more attractive to the increasing number of industrious, patriotic citizens who feared the country was falling into the hands of the plutocracy. Led by William Jennings Bryan, agrarian populist forces succeeded in pushing Congress to pass an income tax law in 1894, only to see it struck down by a conservative Supreme Court. In judging it a "direct"—and therefore unconstitutional—tax, the Court's decision was a plain statement of its opposition not only to an income tax but also to other schemes which arose from "an agrarian and despoiling spirit."

From then on, the income tax was a hot issue in every presidential campaign. Though the tide of public opinion was irresistible, even to the rich themselves, the Court remained obstinately defiant. A constitutional amendment was the only answer. The Thirteenth Amendment, giving Congress the right "to lay and collect taxes on incomes," was passed by Congress in 1913 and ratified less than a year later. Then, with strong backing from President Woodrow Wilson, the Sixty-third Congress carried out its popular mandate by giving the country its first graduated income tax under the new amendment.

Plutocracy

The rule and power of wealth.

Americanization

Immigration and the Process

From its earliest days the United States was a haven for oppressed peoples everywhere and for people anxious to experience the rich promise that was America. They came from many countries and all walks of life, bringing unique talents and cultures that contributed to the diversity and quality of American life. While they were for the most part eager to join the mainstream, a bit of their native culture was inevitably preserved, making America richer. Thus, although there were sporadic outbreaks of nativism (resentment directed against foreigners for their alleged "contamination" of American traditions), most Americans believed that as long as the immigrants were willing to dispense with most of their Old World traditions and to abide by the laws and customs of their new homeland, they were a national asset. On the other hand, the feeling was strong that America had to exercise constant vigilance to keep disruptive influences out of the country (for example, it was felt that foreign nationals who exhibited less than wholehearted allegiance to the American flag should be excluded or deported from the country).

Around the turn of the century, when immigration was at its height and currents of nationalism were strongest, a movement emerged that sought to hasten the assimilation of newcomers into American life. Gathered loosely under the banner of the so-called Americanization movement was an assortment of reformers, educators, and businessmen. Their goal was to strengthen the nation and form a more cohesive, tightly knit society by making citizens out of aliens. There were two roads to take to that end. The first approach, based on a recognition of the immigrants' needs, held sway during the heyday of the Progressive Era, with celebrated reformers like Jane Addams in Chicago and Lillian Wald in New York spearheading the fight against the immigrant's worst enemy—ignorance. Uninformed about the law, the immigrant was an easy prey for unscrupulous employers and merchants. Social workers, operating out of neighborhood settlement houses, worked to prepare the immigrant for full citizenship, so that he could partake of the rights, privileges, and legal protection

Diversity

Unlikeness; complete difference.

Assimilate

To take in and make part of oneself; absorb.

# of Assimilation

that he was entitled to. He was given instruction in the English language, assistance in overcoming bureaucratic obstacles, and vocational guidance. Health and legal centers were established, offering services to their impoverished clientele for little or no cost. This help came without strings attached; no attempt was made to persuade the immigrant to give up his customs for the "superior" American way. Rather, these humanitarian Americanizers, believers in the melting pot, sought to instill in their charges a healthy respect for the cultures of both the New and Old Worlds.

Perhaps because most citizens had little interest in improving the condition of strangers, Americanization was never a very popular cause. But it picked up momentum during and immediately after the First World War when fears were rife that immigrant communities harbored individuals whose loyalty to this country was questionable. Gradually the whole country became caught up in a hysterical campaign to purge foreign influences from American life. The emphasis was no longer on filling the immigrants' needs; the thousands who flocked to the Americanization standard did so hoping to drill conformity into America's foreign element. In evening classes, which were compulsory in some states, the foreign-born were force-fed on civics and American history and bombarded with patriotic leaflets. Although coercion was not widely employed, certain employers adopted the policy of promoting only citizens. Schoolteachers in New York City distributed loyalty pledges for parents' signatures. And in the Revenue Act of 1918 Congress imposed on most aliens an income tax rate double that which citizens paid.

Once peace returned and the "Red Scare" had run its course, the Americanization movement sputtered out. As the country retreated from its global involvement and restrictions were placed on immigration, the threat from abroad seemed to abate. While Americanization had made some achievements in educating the foreign-born, it has generally been thought of by historians as a blot on the American record, a futile, misguided effort.

Clientele

One's customers or clients.

Compulsory

Required; compelled.

Conclude

Reach decisions by reasoning.

After such incidents as the Zimmermann telegram and the resumption of unrestricted submarine warfare, President Woodrow Wilson wearily concluded that the United States must go to war. He harbored no illusions about the war—he knew it would bring untold suffering and privation to the American people—but he felt that he could no longer evade the sobering truth that the German militarists presented a serious threat to the world. It was with a heavy heart that he asked Congress for a declaration of war, but he made it clear that his purpose was righteous. The United States sought no indemnity, no territorial additions, not even the humiliation of the German people. The country was preparing to make monumental sacrifices, the president eloquently declared, "for democracy, for a universal dominion of right by such a concert of free peoples as shall bring peace and safety to all nations and make the world itself at last free." According to Wilson, this "most terrible and disastrous of wars" would be undertaken first to purge and then to reshape the world along benign lines, creating a millennium where war would be unknown. For the moment Germany was the enemy, and all the forces at the disposal of the United States must be marshaled for victory. Those leaders responsible for unleashing such destruction on civilization would receive just retribution. But the German people would be invited to join

Unique

One of its kind; having no equal or like.

the community of peaceful men, bringing their own unique talents to bear for the universal good.

Plans for peace based on these liberal-utopian ideas were underway long before the war had been won. Wilson, as head of a great power, was the hope of idealists and pacifists everywhere. One of their major proposals was the creation of a postwar federation of nations, where differences between nations could be peacefully settled. Wilson took this idea as his own as early as 1916, when he came out in favor of American participation in such an association. In his speech of January 22, 1917, Wilson expressed his belief that the American people were prepared to take on the responsibilities of peace keeping, provided only that it was a "peace without victory,"

War Goals

one which would deal liberally with the German people, replace secret diplomacy with open diplomacy, guarantee freedom of the seas, and give autonomy to oppressed nationalities. The American people, he later said, would not fight for "any selfish aim on the part of any belligerent."

Wilson's proposals were coolly received by the leaders of Britain and France. Three years of bloodshed and sacrifice had left these men determined to wreak vengeance on Germany and to impose a peace so harsh that the country would never again have the capacity to make war. To that end, the Entente powers had drawn up secret plans to carve up German territory and colonies for themselves. They refused to consider Wilson's lenient plan, and the president, who did not want to cause a rupture that might impede victory, decided not to press the issue until the war was won.

The Bolshevik success in Russia on November 7, 1917, forced Wilson's hand. Russia left the war after the Allies refused its request to begin peace negotiations on liberal terms. Condemning the conflict as an imperialist struggle, the Bolsheviks proceeded to publish the heretofore secret documents which revealed that the Allied war aims were as evil as those of Germany.

Wilson had to answer the charges, to set down once and for all the principles for which Americans were fighting. In a speech before Congress on January 8, 1918, Wilson outlined Fourteen Points as a guide for the peace settlement. Many of the points had been made in the "Peace Without Victory" speech: open diplomacy, self-determination, freedom of the seas, and a "general association of nations" to preserve the peace. To these were added the reduction of trade barriers, disarmament, and the fair settlement of colonial claims. The world was thrilled with this noble declaration; it boosted morale and restored a sense of purpose to the many who were exhausted by the years of sacrifice. It gave sustenance to the German moderates, who could now press for peace confident that the fatherland would not be dismembered.

Impose
To put on.

Impede
Obstruct; hinder.

49

Versailles Peace

The Importance of "Peace

Armistice

Temporary peace; truce.

The armistice agreement that was signed on November 11, 1918, contained the stipulation that Germany would have to pay reparations for damages suffered by civilians during the war. This provision, included at the insistence of Britain and France, was a signal of the difficulties that plagued President Woodrow Wilson during the peace treaty negotiations at Versailles. The United States wanted nothing for itself, except an agreement based on Wilson's Fourteen Points; the Europeans, on the other hand, who had made the greatest sacrifices during the war, were determined to exact harsh penalties from the Germans. Although statesmen like British Prime Minister David Lloyd George were largely in accord with Wilson's "peace without victory" position, they were compelled by a vindictive public to squeeze Germany "until the pips squeak." The French gave Wilson a tumultuous welcome in Paris, but at the conference table their leader, Georges Clemenceau, was almost contemptuous of Wilson's idealism. All he and the French people cared about was that Germany be so chastened that it could never again invade French lands.

Wilson knew that his principles would run into stout opposition at Versailles; and, at the height of his prestige as mankind's savior, he was convinced that only through his presence as spokesman for the

Treaty

without Victory"

"peace-loving people of the world" could those principles triumph.

Wilson found that his presence was not enough to overcome the belief that Germany must pay dearly for its crimes. His bargaining power was weak. He had turned the 1918 congressional election into a referendum on his policies, but the Democratic Party was badly beaten at the polls. In Paris it seemed as though "peace without victory" had suffered a popular repudiation, an impression that was reinforced when Theodore Roosevelt, an ardent Wilson-hater, flatly informed the Europeans that because of the election results Wilson had "no authority whatever to speak for the American people at this time." With uncertain public support at home and the European people thirsting for blood, Wilson was forced to compromise on key issues, although the end result of the negotiations upheld many of the Fourteen Points. Despite his displeasure, Wilson went along with French demands for a military occupation of a buffer zone between France and Germany. He allowed the prearmistice agreement compelling Germany to pay civilian reparations—already a staggering sum—to be boosted to account for the costs of pensions for Allied veterans. In violation of the prearmistice agreement and the Fourteen Points, the Allies would occupy the Rhineland until the reparation bill was paid and the French were given ownership of the valuable coal mines in Germany's Saar Valley as compensation for the destruction of French mines by the retreating German armies. Finally the Germans were saddled with a war-guilt clause, making them "legally" responsible for all costs growing out of the war.

Wilson made all these concessions so that he might get what he had long dreamed of—the League of Nations. Despite loud rumblings from home that the Republican-controlled Congress would never approve a treaty that gave outside authority over American troops, Wilson believed that the creation of the League for the preservation of peace was at the heart of the treaty, the provision that would ensure its success. It was on the issue of the League that Wilson would stand, and on this he received his greatest defeat.

Repudiate
Refusal to accept; rejection.

Concession
Anything yielded or conceded.

Invasion of Russia

An Attempt to Stop the Germans

Regime

System of rule or government.

The Russian Revolution in February 1917, whereby the czar was overthrown and a moderate provisional government set up, was welcomed by President Woodrow Wilson. He immediately recognized the new regime and, as a gesture of friendship, dispatched a fact-finding mission to Russia in the spring of 1917, which was supposed to determine Russia's needs for carrying on the war. Had its members done less banqueting and more investigating, they would have learned that pervasive war-weariness, class antagonisms, and popular discontent made Russia a less than reliable ally.

All those factors exploded in the Bolshevik Revolution in November. The new Soviet government immediately concluded an armistice with Germany, one which had a staggering effect on the entire Allied war effort. Released from combat in the East, more than a million German troops would be transferred to the western front. In the hope of convincing the new government to continue to fight and not sign a final peace treaty with Germany, Wilson released the Fourteen Points, intending to show the Bolsheviks that Allied war aims were honorable. Yet the president withheld recognition from the Soviet government. He believed that more democratic elements would soon defeat the Bolsheviks. With the notable exception of Secretary of State Robert Lansing, this misguided belief held sway over most Allied leaders, even after March 3, 1918, when the Soviet leaders signed the Brest-Litovsk peace treaty with Germany.

Predicament

An unpleasant, dangerous or trying situation.

This placed Wilson in a serious predicament. While he abhorred the dictatorial and anticapitalist character of the new regime, intervention in Russia's internal affairs would be a violation of the principle of self-determination. At a time when he was trying to end the war by courting the German moderates with his pledges of a lenient peace, Wilson could hardly act in a manner inconsistent with his own principles without suffering a fatal loss of credibility. But with German reinforcements pouring into the West and with the Bolshevik regime showing no signs of collapse, some action had to be taken to prevent a military calamity for the Allies.

All of the Allies were concerned lest the large quantities of military material stockpiled at Murmansk in northern Russia and at Vladivostok in Siberia fall into German hands. The Bolsheviks were reportedly freeing and arming numbers of German and Austrian prisoners of war. There was also the plight of the Czechoslovak legion, a body of forty-five thousand men that had been fighting the Germans on Russian soil when the revolution took place, trapped in a now-hostile country. A limited military intervention, the Allies declared, would serve to rescue the Czechs, save the supplies, halt the German advance, and possibly bolster the sagging counterrevolutionary, or White Russian, forces.

Wilson reluctantly decided in the summer of 1918, to go along with the demands for American participation in an invasion of Russia. About five thousand U.S. troops, along with twenty-four hundred British and nine hundred French soldiers, were sent to Archangel to protect supplies and the nearby railroad, which was used for evacuation of the Czechs. Nine thousand more Americans landed at Vladivostok alongside seventy-two thousand Japanese. The invasion of Russia accomplished some minor goals, but it failed to undermine the Bolshevik regime or reestablish an eastern front against the Germans.

Bolster

Keep from falling; support.

K.K.K. Revives

Terrorism and New Targets

The Ku Klux Klan, scourge of blacks and carpetbaggers in the South during Reconstruction, arose again in the aftermath of World War I. Its phenomenal growth between 1920 and 1923, when it claimed a national membership of better than three million, was a reflection of the deep sense of discontent which pervaded postwar America. The heady confidence and crusading spirit of the war gave way to the bitter realization that the world was no safer for democracy than before—the American sacrifice of lives and money, it seemed, had been for naught. A serious depression sent the prices of agricultural commodities down; unemployment skyrocketed and credit dried up. At the same time, immigration revived, again bringing to the surface all of the accumulated resentment against foreigners especially when they were Catholic. Not only were the newcomers taking scarce jobs from the native-born, it was believed that they were contributing to the sick atmosphere of lawlessness and immorality spreading over the country by engaging in bootlegging and other criminal activities. In these troubled times, when Americans yearned for a return to the old and proven values of the good old days, the immigrants provided a highly visible and accessible target for their frustrations. The Klan was a perfect vehicle for transforming that discontent into a potent social force.

What the Klan tried to do was to enforce its own brand of law and order over the countryside just as it had done in Reconstruction. At first, blacks bore the brunt of the attacks. After having bravely served their country in the war, many returning black servicemen had expressed their determination to stand up for the civil rights so long denied them. The Klan responded in its own well-rehearsed way. Newspaper headlines told the grim story of Klansmen marauding through black neighborhoods, indiscriminately burning and looting. Blacks were intimidated into working for lower wages and prevented from voting.

But the twentieth-century Ku Klux Klan was different from its counterpart of the 1860s and 1870s. First, its appeal was nearly as

Discontent

Dissatisfaction; uneasy feeling.

Yearn

Desire earnestly.

Intimidate

Influence by fear; frightened.

# of Oppression

wide in the North as it was in the South. This was because after 1920 it began to direct its attention to the purification of white society. For many Americans, that meant ridding the country of the immigrant riffraff and suppressing Catholics and Jews. In the South and Midwest, Catholic officials were hounded from office and replaced with men of the Klan's own choosing. Catholic businesses were subjected to Klan-sponsored boycotts; those who resisted were apt to find blazing crosses on their lawns. Jews received much the same treatment. The KKK also campaigned against indecent pictures and books and fought to end the teaching of evolution in the classroom. White Anglo-Saxons and their culture was to be imposed upon the nation to the exclusion of all "alien" influences.

For a time the Klan became a significant political factor in a number of states including Indiana, Illinois, Ohio, Missouri, Arkansas, Oklahoma, Texas, and Oregon. This foray into politics had a positive effect—the Klan's violence decreased since as its leaders wished to avoid the stigma of association with a terrorist organization.

The Klan, however, ran into stiff resistance. Catholics, black, and Jewish organizations were determined to beat back the Klan. In cities where Irish Catholics were prominent in the police forces Klan members received rough treatment from their law enforcement agents. New York's mayor John Hylan instructed the police to drive the KKK "out of our city." The Klan often had its outdoor initiation ceremonies interrupted by mobs that then pursued and physically assaulted the robed and hooded KKK members.

The Klans achieved some success in upholding Protestant culture, but its larger effect was in mobilizing the opposition. Everywhere Catholics, Jews, and blacks began organizing. Occasionally they even cooperated with one another against the Klan. It was these same groups that would come together in the 1930s to provide the backbone of the new Democratic majority supporting the New Deal—a distinctly untraditional approach to America's problems.

Exclusion

Keeping out; shutting out.

Prohibition

The Official Answer to the

Provoke

Bring about; call forth.

The battle for prohibition was won in a curiously anticlimactical fashion in 1917. For decades this issue had provoked emotionally charged debates between the drys (who charged that the curse of alcohol was sapping the country's vigor) and the wets (who ardently defended their right to drink). It was an issue too controversial for most politicians to touch. But in 1917 by an overwhelming majority Congress quietly passed the Eighteenth Amendment, prohibiting "the manufacture, sale, or transport of intoxicating liquors" in the United States. A little over a year later the last of the required three-fourths of the states ratified the amendment. The Volstead Act, setting up the machinery for enforcing prohibition, followed in a short time. The country was now officially dry—a great experiment in regulating the morals of a whole people had begun.

Why was prohibition passed? For one thing, the ever-increasing dry forces, most of them fanatical in their commitment, were well organized in the Anti-Saloon League. The temperance movement had drawn considerable support from the general reforming spirit of the Progressive Era. And as the drys were making an all-out effort to win congressional support, the First World War had come. The war created an atmosphere of self-sacrifice, moderation, seriousness, and idealism.

But this spirit of reform and idealism soon faded. The war ended in 1918, and servicemen returned home, anxious to resume life and partake of its pleasures. Tired of self-denial, the country turned its energies toward enjoyment. Drinking Americans (a category that included most citizens) meant to have their liquor, whether the law allowed it or not. People began buying and operating illegal home stills, turning out gallons of corn or "moonshine" liquor. City residents without a domestic source frequented speakeasies—saloons operating on the sly—or bought bottles of what looked like whiskey from door-to-door bootleggers. All too often these beverages turned out to be one of several unwholesome or lethal substances that were being palmed off on unsuspecting customers—Jamaica Ginger, for

Provoke

Bring about; call forth.

Illegal

Against the law.

"Curse of Alcohol"

example, which paralyzed thousands, or Yack Yack Bourbon, a blend of iodine and burnt sugar.

From the start organized crime was deeply involved in the business of quenching the nation's thirst. The profits were enormous and the risks minimal. The government's force of prohibition agents—never more than three thousand ill-paid, overworked men—was far from adequate for the tasks of patrolling all the waterways for rum-runners and of locating and closing all the stills and speakeasies. Their predicament was compounded by the fact that the public often sided with the lawbreakers. Gangsters like Al "Scarface" Capone, who controlled nearly all the illicit alcohol in Chicago, were commonly thought of as glamorous celebrities. Many thoughtful citizens became concerned that the widespread disregard for prohibition would breed contempt for law in general. In 1933 the noble experiment, its failure acknowledged even by its supporters, was repealed by the Twenty-first Amendment.

Minimal
Smallest; least possible.

59

The U.S. Occupation

Making the Caribbean Safe

The election of Woodrow Wilson in 1912 seemed to promise a new era in relations between the United States and the Caribbean republics. If the pronouncements of the new administration could be believed, American capital would no longer be used as an instrument of political control over foreign countries. "One of the chief objects of my administration," Wilson declared a week after taking office, "will be to cultivate the friendship and deserve the confidence of our sister republics of Central and South America." He pledged himself to a policy of self-determination and respect for the sovereignty of other nations.

But despite Wilson's pledges, his Caribbean policies revealed little that was new. The U.S. employed massive military intervention, economic and political intimidation, and arrogant paternalism. On more occasions than his two "imperialist" predecessors combined, Wilson sent marines into supposedly troubled lands. And though he spoke of cultivating friendship, he also made it plain that the Latin Americans would have to behave themselves in a civilized fashion to be eligible for such evenhanded treatment. Both he and Bryan were sure that they knew what was best for other peoples. While Wilson was to a certain extent obliged to pursue certain established objectives, such as protecting the Panama Canal area and safeguarding American capital, the principle motive behind his interventions was, as one historian has put it, "the promotion of democracy, or at least its negative side, the prevention of a chronic state of revolution." Often these dual purposes were served nicely by interventionist policy. Suppressing revolutionary activity would not only give rise to a climate where democracy (and American capital) could flourish, but would also eliminate a possible pretext for foreign (i.e., European) intervention. All of these elements were present in Wilson's policy toward Haiti.

The situation in Haiti in 1914 was nothing short of chaotic. Ninety-five percent of the population were impoverished and illiterate. Epidemics swept through the ramshackle peasant dwellings with

Cultivate

Develop; improve.

Pursue

Follow in action.

Illiterate

Not being able to read and write.

60

# of Haiti

for America

deadly regularity. The treasury was bare, looted by corrupt politicians and drained by years of revolution. Creditors were going unpaid, and the political system was a nightmare. Rival politicians drew support not from the people, but from bands of mercenaries known as *cacos*. It seemed that no sooner was one president placed in office (usually by force of arms) than another pretender gathered up a caco army and headed for the capital. Elections were a mockery.

Wilson and Bryan took upon themselves the monumental task of putting this sad house in order. In the beginning of 1914, they began formulating plans to establish a customs receivership over Haiti. Under that scheme the United States would superintend all of Haiti's financial affairs, from collecting and allocating revenues received from customs duties (which were the principle source of national income) to completely reorganizing the economy. Haiti would not be permitted to increase the size of its national debt without first obtaining American consent. Additionally, the Americans would be allowed to establish a naval base. Order would be maintained through an American-trained- and -controlled national police force, with the backing of U.S. marines if necessary. A team of American engineers would be appointed to supervise public works and to achieve higher standards of sanitation. In short, this scheme would make Haiti a virtual protectorate of the United States.

At first, Wilson and Bryan were reluctant to force this treaty on the Haitians. Then, in June, 1915, an event occurred in the capital of Port-au-Prince that drained Wilson's last ounce of patience. The current strongman, General Vilbrun G. Sam, was fighting off another group led by Dr. Rosalvo Bobo. As a warning to the insurgents, Sam executed sixty political prisoners, including a former president, Oreste Zamor. The city's people rose up in spontaneous anger, dragged Sam from his quarters and hacked his body to pieces. Wilson sent in the marines on July 25. A pro-American politician, Philippe Dartiguenave, was installed as president, and in September he signed a treaty with the U.S., authorizing its intervention in Haiti.

Insurgent

A person who rises in revolt.

Immigration

Fear and Loathing among

Congenial

Compatible; agreeable.

It must be remembered that the nativist sentiment of the Twenties was not directed against all immigrants. Even the most rabid nativists were willing to admit as many Britons and Scandinavians as desired entry, because they were of a racial and cultural cast congenial to most native-born Americans. The objects of the nativists' fear and loathing were the Southern and Eastern Europeans—the "unwashed multitude" of Slavs, Poles, Greeks, Italians, and Jews. The nativists viewed them as pouring into Ellis Island and carrying the contagion of radicalism and disease. The melting pot bubbled incessantly, but the number of daily arrivals severely taxed its capacity to blend them into a uniform society. Nativists cried that unless something were done to limit the flow, the cherished American institutions and way of life would become infected, wither, and die. The nativists wanted legislation that would put an end finally and inalterably to the policy of unrestricted, unlimited immigration.

Debase

To lower quality or value.

The idea of immigration restriction was not new; it had been proposed as early as 1885, but since the problem had not yet assumed significant proportions in the popular mind, it did not make much headway. Many held that abandonment of refugees would debase the American ideal and depart from the liberal traditions that had made this country so special. Big business counted on cheap immigrant labor for its work force, so it strongly objected to imposing limitations. By 1920, however, immigration restriction was an idea whose time had clearly come. It was no longer just the goal of a few noisy extremists; many decent Americans cast aside their egalitarian scruples and eagerly embraced restriction as a solution for the nation's many problems. Organized labor, jealously guarding its recently won gains, supported the effort. And big business, affected by increasing automation and the movement of blacks north from the cotton belt, no longer had a pressing need for an abundant supply of immigrant labor. The only voices raised against the pending legislation were those of the foreign-born themselves.

Restriction

the Native-born

In 1921 a bill passed by the House that provided for a complete, one-year suspension of immigration was defeated by the Senate, which substituted its own measure. It set down a policy on immigration that, in its broad outlines, remains in effect to this day. It also expressed and passed into law the national propensity to discriminate against those considered religious and ethnic inferiors. That bill, signed into law by President Warren G. Harding in 1922, established quotas for each nationality, with exceptions for humanitarian and diplomatic needs. (The recent influx of Vietnamese to this country is a good example of the system's flexibility.) America's open-door attitude toward immigrants, a policy which accounted for one of the greatest movements of people in the history of the world, had thus come to an end. And it ended not in a straightforward manner but in a way that inflated America's prejudices against certain outsiders. In recent years the U.S. has changed its immigration laws to eliminate discriminatory intent but not its restrictionist stance.

Suspend

Exclude or remove for a while from some job or privilege.

63

Boston Police Strike

The Debate over the Right to Strike

In 1919 the Bolshevik peril loomed large in the minds of Americans. It was a time when labor unrest took on a new gravity and when each outbreak added to the public's anxiety. Were the militant workers loyal to their country, or to some alien creed? Were we heading toward the chaos of class warfare? The only available evidence seemed to confirm the worst of answers. Socialism was sweeping over Europe and the plague was already lapping at our shores. American radicals, inspired by the Russian example, were loudly proclaiming the need for revolution. Business leaders viewed this situation as an opportunity to discredit unionism once and for all and described in frightening detail the fate that would befall the nation if it failed to nip revolution in the bud. They warned that any victory for labor, no matter how small, was a victory for those who sought to destroy the American way; submission to labor's demand meant certain demise of the capitalist order.

The public, which trusted the word of business, was alarmed at what its leaders were now saying. Each confrontation with unions and workers became an acid test of the country's will to preserve its institutions. The fact that labor had legitimate grievances became a secondary matter. In that framework, there was no such thing as overreaction; the stronger the "American" response to disorder, the better. In case after case, national guardsmen and police were used to quell labor disturbances. To those on the side of the American system, this use of official force was necessary so that decent society might be saved from anarchy.

But was society safe? The possibility that the police themselves would abandon it and join the ranks of labor's malcontents was too dreadful for most Americans to contemplate. The very idea of a police union seemed somehow a contradiction in terms, as if a policeman could not be active in his own behalf and serve the public interest at the same time. That was the opinion of Boston Police Commissioner Edward U. Curtis, a conservative lawyer-banker with a long record of hostility to labor. Curtis had continually refused to negotiate with his

men over their demands for higher wages, shorter hours, and improved working conditions. When the frustrated police applied to the American Federation of Labor for affiliation, Curtis responded with an order prohibiting his men from joining any group that was other than fraternal. Undaunted, the police went ahead, receiving an A.F. of L. charter on August 15, 1919, and pledges of support from Boston's Central Labor Union. Curtis, seeing his authority so boldly challenged, dismissed from the force nineteen leaders of the new police union. Outraged at the commissioner's arbitrary and repressive actions, the union membership overwhelmingly passed a motion to walk off their jobs. On Tuesday, September 9, Boston found itself unprotected.

On the first night of the strike, hoodlums ran brazenly through the city's streets, looting and rioting as police stood by. Passers-by were molested and several shootings took place. Though the actual amount of damage done was small, newspaper accounts so exaggerated the violence as to leave no doubt the city was entirely in the hands of Bolshevik mobs. More determined than ever to break the strike, Curtis called on Governor Calvin Coolidge for state militia to patrol the streets. At the same time, civilians were armed and organized into a citizen protection force. By September 11, with the Governor assuming personal command, the city was well under control.

The policemen quickly realized that the strike had turned into a colossal blunder. The police were willing to submit their grievances to arbitration and return to work, but Coolidge and Curtis would have nothing of it. The commissioner was bent on teaching them a lesson: not a single striker would be rehired; an entirely new force would be recruited. Coolidge backed that decision. "There is no right," the Governor said, "to strike against the public safety by anybody, anywhere, any time." The words made him a hero to a nation that was looking for someone to stand up to the presumed threat of labor insurrection.

Brazenly
Shamelessly.

KDKA: First Radio

A New Form of Mass Entertainment

After years of experimentation, wireless broadcasting was finally brought into practical existence by Lee De Forest's invention of the audion tube in 1906. The early pioneers of radio technology saw it as a substitute for the telephone and the telegraph. They viewed the radio, however, as having more personal and private applications. The fact that anyone with a set could, and did, listen in to transmitted messages opened the door to other possibilities. De Forest himself looked forward to the day when the "news, and even advertising, will be sent out to the public on wireless telephone." To prove and publicize radio's utility, the Westinghouse Company, a leader in the infant industry of manufacturing radios, set up the first commercial station, Station KDKA in Pittsburgh, Pennsylvania, which began transmitting on the election night of November 2, 1920. This first scheduled radio broadcast sent returns of the Harding-Cox presidential race into the homes of the few hobbyists who had receiving equipment.

Americans were impressed by this initial demonstration, but radio was as yet a novelty, with limited programming and fewer auditors. By 1922, however, the results of David Sarnoff's farsighted imagination in the young industry began to pay off, convincing skeptics that radio had unique entertainment value. As he rose in the ranks of the radio business, Sarnoff was a constant booster of radio's potential as a "household utility in the same sense as the piano and the phonograph." As commercial manager of the Radio Corporation of America, in 1920 Sarnoff urged the establishment of a network of commercial stations offering a variety of music, comedy, news, drama, and sports. Two years later there were four hundred stations in regular operation all over the country, and the demand for receivers, even at $275 each, was such that the major manufacturers went to work in round-the-clock shifts. That year there were radios in over three million homes; the sale of sets mushroomed into a $60 million industry. In 1923 alone the General Electric Company sold eleven million receivers, with no end in sight to the demand.

Skeptic
A doubter.

66

Broadcast

Captivates the Nation

Radio burst on the scene with explosive impact. In a country starved for entertainment, the radio was manna. After the initial outlay, one could attend the theater, concerts, sporting events, even church services without having to leave the comforts of home. Social commentators and clergymen gave radio their blessing as a cohesive force in family life. Millions of American families, brought together for their favorite programs, sat in front of their sets, listening intently to the Ipana Troubadors, the warbles of Rudy Vallee, or Grantland Rice's broadcast of the World Series. A whole new lexicon of radio jargon was born, and hitherto unmechanical men talked endlessly of ways to eliminate static and the relative merits of different brands of receivers. President Warren Harding had an outfit installed in his White House study, and politicians delivered their addresses into microphones, sending their words into many more homes than could be reached otherwise. Radio, as much as anything else the decade produced, dramatically altered the habits of Americans.

Eliminate

Leave out.

EIGHT WHITE S
ON CHARGE OF
CICOTTE GOT

enkee Of ers Give Pre
A d Offer Him

How Corruption Tarnished the

One of the most appealing aspects of baseball when it emerged as the national pastime around the turn of the century was its uprightness. It was a notably "clean, straight game" played by "ruggedly honest" athletes. The competition on the ballfield, therefore, provided a fitting example for the nation's youth and a respite from the distressing realities of war and corruption for their elders. While boxing, wrestling, horse racing, high finance, and politics (among the other "sports" favored by Americans) seemed constantly tainted by scandal, baseball, so its boosters believed, was incorruptible. How could you fix the outcome of an unpredictable event participated in by eighteen men? To Americans ballplayers were somehow a breed apart—clean-living demigods who would never betray the trust placed in them.

Yet those close to the game could have told them that baseball was not as pure as it seemed. Fans were not above placing occasional small wagers on the outcome of the game or participating in weekly betting "pools." Betting went on in the ballparks themselves and sections of certain stadiums became notorious as gambler hangouts. But no one seemed to feel that there was anything wrong with casual gambling on baseball, provided the games were honest—and everyone assumed they were. Nonetheless with large sums of money wagered on each game the temptation to control the outcome was great and unscrupulous elements were prepared to try any ploy, legal or not, to work the odds in their favor. Gamblers gradually learned that certain players, who were dissatisfied with their salaries, could be bribed to "throw" games. Club-owners and the league officers were aware of this but chose to hush up the matter to avoid embarassing revelations.

That is why revelations of scandal in the 1919 World Series created such a sensation. Rumors that several members of the Chicago White Sox team had been paid to lose the series to the underdog Cincinnati Reds had begun to fly even before the series was over. The owners responded to the allegations with righteous indignation. Charles

Taint

Corruption; disgrace.

Unscrupulous

Without principles; not careful about right and wrong.

PLAYERS ARE INDICTED
IXING 1919 WORLD SERIES;
$10,000 AND JACKSON $5,000

to Comiskey
of Their Whole Team

COMISKEY SUSPENDS THEM

Promises to Run Them
Out of Baseball if

Great American Pastime

Comiskey, owner of the White Sox, offered $10,000 to anyone who had proof of wrongdoing on the part of "his boys." This was entirely out of character for Comiskey, who had a reputation as one of the most miserly of baseball men. But no one came forward with anything conclusive, and with the 1920 season about to begin, the matter soon dropped from the public notice. It remained forgotten until September 4, 1920. Then the news broke that a National League game between Chicago and Philadelphia had been thrown by certain Chicago Cub players. There was enough evidence to warrant investigation by a Cook County grand jury into the whole question of gambler involvement in professional baseball. It soon became clear that the Chicago-Philadelphia fix was only the tip of the iceberg. Known gamblers were subpoenaed, and the stories they told implicated a number of major leaguers, among them eight Chicago White Sox players: star pitcher Eddie Cicotte, first baseman Chick Gandil, shortstop Swede Risberg, batting champion "Shoeless" Joe Jackson, and four lesser lights. Testimony revealed that the men had been promised up to $25,000 each to make sure that the Cincinnati team won the series of 1919, though none of them ever received more than $5,000. On the witness stand, Joe Jackson told of moving slowly after balls hit to him, making errant throws, and deliberately striking out with men on base. "Say it ain't so, Joe," begged a disbelieving boy of his hero outside the courthouse. "I'm afraid it is, son," Jackson answered ruefully. The team came to be known as the "Black Sox" and the eight fixers were permanently suspended from ever playing professional baseball again. The scandal had enough impact to push the Red Scare off the front pages of the nation's newspapers. The media called for immediate action in rooting corruption from baseball. The owners realized that crookedness, where it existed, could no longer be swept under the carpet. To the end of policing themselves, team-owners appointed the stern Judge Kenesaw Mountain Landis, famous for prosecuting political radicals, as baseball's new commissioner.

Reputation

What people think and say about the character of a person.

Errant

Roving; wandering.

69

Women's Suffrage

American Women Win the Right

> *Resolved,* That it is the duty of the women of this country to secure to themselves their sacred rights to the elective franchise.
> Ninth Resolution of *The Declaration of Sentiments,* adopted in July 1848 at the Women's Rights Convention in Seneca Falls, New York.

Resolution
Something you are determined to do or have decided to do.

This resolution, offered by Elizabeth Cady Stanton to the assembled delegates, was the only one of twelve not unanimously adopted. Even among the leaders of the women's movement, suffrage was an idea that provoked dissension. Conservative feminists—they preferred to call themselves practical—chided their more militant counterparts for including in the declaration a measure that stood virtually no chance of being accepted by male America. They argued that insistence on immediate suffrage would alienate even those men who were sympathetic to the cause of women's rights. (There had been some limited examples of women voting before in the United States. In Europe most men couldn't even vote at this time.) It would be far better, they said, to make gradual progress toward winning men over to their views than to stake all their hopes on a then-impossible dream. Other women believed that the vote was essentially unrelated to their needs. They reasoned that involvement in the grubby world of politics would only serve to corrupt the idealism that women were hoping to bring into practice. They wanted to be free of the restraints imposed by law on their activity, but only so that they could work from the ground up through social work and moral instruction.

Grubby
Dirty; rotten.

As many predicted, these first stirrings of organized feminism achieved no notable breakthroughs. Men perceived this movement as a threat to the country's whole social structure, which was based on distinct sexual roles. Men knew that with the vote would come demands for women to have the freedom to plot their own courses in life. Women might displace men in business and industry. Worse

to Vote

still, the family unit, traditionally the foundation of American society, would suffer if women pursued individual careers.

The National American Woman Suffrage Association, a union of two smaller groups, was founded in 1890. The suffragettes decided to concentrate their efforts on winning over individual states, one by one, rather than work on the national level. This was no small task. Everywhere they went the suffragettes found apathy and the well-financed opposition of the liquor and brewing interests, who naturally assumed that enfranchised women would go to the forefront of the prohibition movement. In the face of this hostility, women had won only four states by 1896—Colorado, Idaho, Wyoming, and Utah.

Almost everyone expected that the advent of women's suffrage would leave the American system profoundly altered, for better or worse. Yet the outcome of the experiment in those four states, as well as in the states that followed, was surprising. Women were neither voting as a bloc nor infusing politics with a "feminine influence." Once it became clear that the vast majority of women had nothing more drastic in mind than to have a voice in government, more and more state legislatures, especially in the West, where sexual roles had always been less clearly defined, bestowed the franchise on their women. By 1918 women had complete suffrage in eleven states and the right to vote for the president in twenty-three others.

Thus when the suffragettes began the campaign that culminated in the adoption of the Nineteenth Amendment, many of the battles had already been won. However, they still encountered stiff resistance in the southeastern states, and Congress was reluctant to go along. But these opponents were no match for the suffragettes. President Wilson came out formally on the side of the women's cause in 1917. The same year the first woman ever to sit in Congress took her seat—Jeannette Rankin of Montana. Two years later Congress passed the Nineteenth Amendment to the Constitution, which declared that "the right of citizens . . . to vote shall not be denied . . . on account of sex."

PRESIDENT
...his is the time to s...

Teapot Dome

Scandal in the Oil Fields of Wyoming

In 1909, the government began to set aside three tracts of oil-bearing land for the use of the navy in the event of an emergency: Naval Reserve No. 1, at Elk Hills, California; No. 2, at Buena Vista, California; and No.3, at Teapot Dome, Wyoming. It was generally thought to be an altogether prudent precaution, applauded by navy men and conservationists. The oil companies, however, grumbled over their exclusion. In the twenties, however, business interests were in the saddle again, encouraged by Washington administrations willing to defer to private enterprise. In the Harding administration in particular, they also found such a relaxed moral climate that it became possible to purchase generous concessions from well-placed individuals. Bribery was commonplace in the cutthroat competition of the business world; but never before had it entered so flagrantly into the chambers of government.

The central figure in the Teapot Dome scandal was Secretary of the Interior Albert B. Fall, formerly a senator from New Mexico. He was

Prudent

Carefully planned ahead of time.

an avowed anticonservationist willing where circumstances permitted to turn government lands over to his friends in the private sector for development. Shrewd and aggressive, he managed to convince his weak-willed colleague, Navy Secretary Charles Denby, to transfer jurisdiction over the government oil fields to the Interior Department. This done, Fall next went to work on President Harding, informing him that oil in the government fields was draining into the private land surrounding them. Although we now know that Fall greatly exaggerated the problem, Harding was receptive to Fall's proposal that the government sell what oil was still available before it disappeared entirely. On April 7, 1922, Fall secretly and without competitive bidding leased the Teapot Dome Reserve to Harry F. Sinclair's Mammoth Oil Company and the Elk Hills Reserve to Edward F. Doheny's Pan-American Oil Company in December.

It is likely that these transactions would have gone unnoticed had Fall not suddenly begun to live the life of a man who made more than the $12,000 salary of a cabinet officer. (Fall, for example, was seen paying off his creditors with new $100 bills.) Rumors began to fly that Fall had manipulated the leasing deals for his own enrichment. A Senate Committee, headed by Thomas Walsh of Montana, started an investigation. It was revealed that Fall had indeed profited from the sale—to the tune of $425,000. Doheny's son had delivered to Fall at the time of the deal a black bag containing $100,000 in cash. Sinclair transferred $233,000 in Liberty Bonds to Fall's son-in-law, in addition to $85,000 in cash and a herd of prime cattle. Fall was indicted and convicted on charges of bribery, and he received a $100,000 fine and a one-year jail term—the first cabinet officer ever to go to prison. Sinclair and Doheny got off lightly: the leases were voided by the Supreme Court, and Sinclair had to serve a six-month prison term for contempt of Congress and for jury tampering. Ever since, Teapot Dome has come to symbolize flagrant abuse of office by a public official. (Watergate has since replaced Teapot Dome for this dubious distinction.)

Shrewd
Clever; having a sharp mind.

Rumor
Statement or story that is thought of as true without any fact to prove it.

Calvin Coolidge

A Businessman's President

News of President Warren Harding's death reached the little town of Plymouth Notch, Vermont, at 2:45 A.M. on August 3, 1923. As kerosene lamps glowed in the Coolidge house, in the sitting room the justice of the peace began reading the presidential oath of office to his son Calvin. The ceremony lasted about two minutes, whereupon the father and the son went back to sleep.

This quaint scene set the tone for the administration of Calvin Coolidge, thirtieth president of the United States. The Puritan simplicity, rugged honesty, and dry frugality which Coolidge brought to the White House stood in contrast to the self-conscious, high-spirited, wild-living temper of the twenties, but Coolidge's character qualities endeared him to the American people. Coolidge offered reassurance that the old virtues, enthroned in the nation's highest office, could comfortably exist alongside the new ones and even complement the life-style of prosperity. The president, true to his Yankee upbringing, might have been austere in his personal habits, but the tenor of his administration was to encourage, rather than to condemn, conspicuous public consumption.

Coolidge was the perfect man for the times, a refined version of his flawed predecessor whose mediocrity had nonetheless struck a responsive public chord . While Harding's platitudes made him seem like an amiable boob, Coolidge's succinct wit captivated. Harding had come to office by way of the smoke-filled back room and political infighting, Coolidge by popular acclaim rising out of his stand during the Boston police strike. Shy, taciturn, even dour (Alice Roosevelt Longworth said he looked as if he had been weaned on a pickle), he was the most unlikely of candidates for idolatry, but in the twenties Americans set different standards for their heroes. In 1924 the slogan Keep Cool with Coolidge was more than just campaign rhetoric; the president himself seemed wholly unflappable and under his rein the country might enjoy similar stability. Coolidge was a man of convictions, who brought with him some definite ideas on how the government should be run. One author wrote that Coolidge "aspired to

Quaint

Odd or strange in a pleasing, interesting way.

Austere

Harsh; stern.

in the White House

become the least president the country ever had; he attained his desire." He vehemently opposed federal spending and restraints on business, exercising his veto power fifty times over such measures as the farm relief bill and pay and pension increases for federal workers and veterans. Imbued with a deep respect for the business ethic, he deliberately instituted a businessman's government at a time when that breed was held in the highest esteem. Whereas Harding's reputation had been injured by the misconduct of his underlings, Coolidge, although he did not harp on the shortcomings of his predecessor, quietly set his official house in order, making conservatism respectable. Coolidge enjoyed his office and took a childlike delight in the trappings of power. He used to press all the buttons on the intercom in his oval office to see "his staff" scurry in to do his bidding. By 1927 the novelty had passed, however, and Coolidge put an end to the game in characteristic fashion. "I do not choose to run for president in 1928," he said, and that was that.

Hollywood

The Movie Boom and the Move

In 1909, Hollywood, California, was a sleepy little suburb of Los Angeles. Then the movies came. Almost overnight Hollywood was transformed into the film capital of the world, the showcase for some of its greatest talent, and the most glamorous and notorious city in the country. The attention of millions of Americans was focused upon life in L.A. From afar people followed the exploits of stars like Douglas Fairbanks, Charlie Chaplin, Mary Pickford, Clara Bow, and Theda Bara and were titillated by the exclusive parties, the sin, the scandal, and the debauchery as it existed in movieland. Yet all the sensation was subordinate to Hollywood's highly visible industry: the multimillion dollar business of making motion pictures. By the midtwenties the film culture had so saturated Hollywood that the city and its industry had become synonymous.

There was nothing accidental about Hollywood's rise to urban celebrity. If nature had, in consultation with the film people, set aside one place in the country for the pursuit of cinematic craft, Hollywood would have been it. Hollywood was endowed with certain physical traits that made it a veritable Promised Land for the industry. It had a superb climate and was located adjacent to a major metropolis, yet land (needed in abundance for studios and sets) was still relatively cheap. Nearby were the mountains, deserts, seashore, and farm lands so essential for location work. To this attractive package were added man-made inducements. The California legislators offered generous tax concessions and lenient labor laws.

Still, the industry was slow to capitalize on these advantageous circumstances. In the early days, New York City was the center of film-making. There, producers could draw from the legitimate theatre for a steady supply of notable actors and actresses (it having already been determined that business at the movie theater box office ran in direct proportion to the illustriousness of a film's performers). In New York too dwelled wealthy investors, without whom no major film could be financed, for film-makers, responding to the public's demand for more lavish screen displays and bigger names,

Notorious
Famous; well-known.

Veritable
Actual; real.

Lavish
Extravagent.

were spending more—much more—on their productions. The typical two-reel subject that had cost from $500 to $1,000 in 1912 was in 1915 a five-reel feature costing up to $20,000, with "spectaculars" going even higher. With such sums on the line, nothing could be left to chance. Directors, who had once worked at a leisurely pace, now had to adhere to strict shooting schedules. And every day lost to New York's unpredictable weather meant added and unnecessary expense. During the winter all activity was forced to a stop. As early as 1910 some of the Eastern companies began to look elsewhere for a more congenial setting.

And so by 1916, Hollywood was on its way to eclipsing New York as the film capital of the world. New York and New Jersey were deserted as entire companies were moved west. At first there was some resistance from the Hollywood natives who feared that the movie business would turn their peaceful community into a present-day Sodom and Gomorrah. Some local hotels displayed signs reading "No dogs or actors." But, for the most part, the film establishment was well received. Few natives could resist what was tantamount to instant prosperity.

Indeed, for the first few years, Hollywood grew at such a phenomenal rate that housing and public services were at a premium. Apart from the major studios and all their employees, hundreds of screen hopefuls flocked to Hollywood, Mecca for the aspiring actor or actress. Gossip columnists floated around the city, uncovering (or fabricating) scandal for the delectation of readers all over the country, leading them to believe that, in Hollywood, dissipation and materialism reigned.

True or not, this exposure helped the movie business flourish. People were attending cinemas in unprecedented numbers—in 1929 ninety-five million tickets were sold. Studio stocks were listed on Wall Street, and movie-making was the country's fifth largest industry. The movies had put Hollywood on the map and the movies themselves were the prime beneficiary.

Tantamount

Equivalent in value, effect or force.

Arms Control

Disarmament in the Postwar Years

The war was ended. Americans paused in their celebration of the armistice to assess the damage done and the prospects for the future. Contemplation of the devastation of life and property gave rise to an inescapable conclusion: another war was unthinkable. What the United States would do to further the cause of world peace was the question on everyone's mind.

The repudiation of the Treaty of Versailles and the League of Nations demonstrated the depth of the feeling that the European powers were, at the heart of it, wholly responsible for the war. For this reason the United States should avoid political entanglements abroad, since to do otherwise might compromise America's independence of action if the Europeans should go at it again.

Although Americans were on the whole untroubled by the government's inaction, they always paid lip service to the cause of peace. The country was in no mood to commit itself to long-term help to ailing Europe. However, the United States was interested in ways of curing the world of its ills, with minimum involvement. Nearly everyone had his own theory on how to put an end to war. These were taken to their illogical conclusion in 1927 when a professor at Columbia University, James T. Shotwell, came up with the idea that war, like drunken driving, could be outlawed. This led to the Kellogg-Briand Pact of 1928, which was signed by most nations, including Japan as it was preparing forces for an invasion of China.

One of the popular panaceas of the day was disarmament. If the great powers rid themselves of the implements of destruction, there would be nothing for them to fight with. Disarmament was attractive for other reasons as well. Instead of the government wasting huge sums on battleships, it could use the money for more constructive social ends. Republican Senator William E. Borah of Idaho, the leading proponent of disarmament in the United States, thought it was worth a try. He urged President Warren Harding to invite the representatives of Great Britain and Japan (those from France, Italy, China, Belgium, Portugal, and the Netherlands were added later) to

Devastate

Destroy; lay waste; unfit to live in.

Panacea

A solution to any problem.

a conference on the reduction of "naval expenditures and building programs." The public, of course, responded favorably to the proposal. Leading journals such as the Chicago *Tribune*, the New York *World*, and the San Francisco *Chronicle* came out strongly in favor of disarmament. Civic and pacifist groups joined them in supporting the limitation of armaments.

The conference convened in Washington on November 12, 1921. After a rousing speech of welcome from President Harding, the delegates settled back to hear what they assumed would be a general keynote speech by Secretary of State Charles Evans Hughes. Instead, Hughes delivered an electrifying address which specified the reductions each of the three major naval powers (the United States, Britain, and Japan) would be called upon to make. Altogether he proposed that no fewer than sixty-six capital ships be scrapped. Furthermore, his plan provided for a ten-year moratorium on the building of warships, as well as a permanent ratio which would give Japan a fleet forty percent less than that of the United States and Britain. Italy and France received proportionately lower allotments. After Japan forced some minor adjustments in its favor, the conference adopted the plan in total. For the first time in history, major powers had actually consented to disarm.

The agreement, however, was seriously flawed in several respects. First, it mainly covered battleships and heavy cruisers, which most countries were in the process of phasing out in favor of lighter, more maneuverable craft. The pact did not mention submarines, which would play such a decisive role in World War II, or land armaments and air power. As far as it went, the treaty was fine; for the time being another costly naval race had been avoided, and it brought a relaxation of tensions in the Pacific. But the treaty was doomed to failure and the American people attached unreasonable expectations to it. It did not bring universal peace, as they had hoped, because there were forces at work in the world that did not lend themselves to easily negotiated settlements.

Armament
War supplies and equipment.

Flaw
Fault; slight defect.

Road Building

Automobile Travel Made Possible

The automobile began to come into its own just before World War I. People were buying cars only to find that, outside of the East, there were almost no paved highways suitable for motor travel. State legislators maintained the car was a pleasure vehicle and allocated little for new road construction. And local townships and counties were unable to keep up with the demands for improvement. The manufacturers and the motoring public quickly realized that only the federal government had the resources to build an extensive and uniform system of highways.

The first step was taken in 1916 with the passage of the Federal Highways Act. Federal funds would now be offered to match every dollar spent by state and local authorities on road building. The program had hardly gotten off the ground when the war brought it to a halt. Under the wartime strain, however, traffic increased so much that by 1918 it was estimated that half a million motor trucks and five million motor cars were in use on the nation's already-overburdened highways. When peace returned, road building was placed near the top of the country's priorities. The federal government renewed its earlier contributory fund, and the states hastened to comply with its requirements. By 1919 nearly forty states had organized highway departments. The urgency had become so great that there was only scattered opposition to federal highway aid, mostly from the railroads. The War Department saw improved roads as crucial for the defense or mobilization of the country. And the Agriculture Department gave road building all the encouragement it could muster in the belief that the new roads would speed the flow of farm products to the market.

By the early twenties the program was in full swing, though problems persisted. Some states were unable to commit enough capital to qualify for federal matching funds, while others embarked on ruinous schemes of deficit financing. Many welcomed federal aid, but were not pleased with the regulations. In some areas shoddy materials and workmanship caused roads to crumble under heavy traffic.

Extensive

Large; far-reaching.

Priority

Coming before in order of importance.

82

The welter of new roads, many either unnamed or with destinations unclear to motorists, caused confusion. And, as Detroit produced speedier cars, some highways designed for travel at a more leisurely pace were obsolete before they were finished. Wider lanes were needed for the nine million automobiles on the roads by 1920.

Welter

Confusion; commotion.

Gradually these problems were ironed out. Most states found that the gasoline tax, which placed the cost of construction squarely upon highway users, was a fair source of revenue. Federal inspection of work in progress was stepped up, ensuring qualitative uniformity. In 1925 the secretary of agriculture approved a standardized system of numbering and marking highways. Meanwhile, engineers worked hard to upgrade roads which had been proven unsafe or inadequate.

By 1921 total expenditures for road building from all sources had reached nearly a half billion dollars annually with the federal government supplying about 40 percent of this amount. Seven years later that figure had climbed to more than a billion dollars spent annually on highway construction and maintenance. In the first decade of federal aid to highways almost 56,000 miles of road had been completed. By 1928 the United States could boast of one of the world's finest systems of well-surfaced roads.

Billy Mitchell

The Move to Establish an Air Force

Musket

A kind of gun which was used before rifles were invented.

Brigadier General William "Billy" Mitchell was a man with a mission. He dreamed of the day when the airplane would be accepted for what he believed it was—the most revolutionary implement of war since the musket, a weapon that would profoundly change the nature of modern warfare.

For centuries warfare had been regarded as a science, no less (and often more) respectable than optics or astronomy. Battles typically took place on open fields, where the opposing riflemen and artillerymen could best see each other. Everything followed the book. Thrust was met by counterthrust. And so it went, each move cool and methodical, until one side had had enough or until night fell, when both armies retreated, to pick up the battle another day.

For the majority of citizens behind the lines, life was virtually untouched by the disciplined violence. Often a holiday atmosphere swept a country that was going to war. Civilians who protested were usually prompted less by humanitarian considerations than by resentment over taxes raised to cover the costs of maintaining an army. Soldiers heading to do battle in 1914 received the same rousing send-off as had countless generations of warriors before them. The civilians settled back into their routines, confidently expecting a speedy and glorious end to the fighting. But before too long it became clear that something was terribly different about this war. This was no armed squabble between two mercurial princes; it was a life-and-death struggle between two sides equally resolved not to give an inch. From the start, the bloodshed at the chaotic front lines was staggering. Men fell in droves before the immense destructive power of such newly developed weapons as the machine gun and the tank. One novel device, potentially the most awful, was the airplane. It appeared in combat for the first time in the First World War. The dogfights that broke out over the European battlefields were among the war's more interesting spectacles, and although never deployed strategically, the airplane was profitably used on many reconnaissance missions. This did not satisfy men like Billy Mitchell. Mitchell,

Squabble

A noisy, petty quarrel.

84

who was General Pershing's leading combat air commander, fervently believed that the airplane had unlimited possibilities as an offensive weapon, possibilities that were being ignored by shortsighted commanders. He proposed that an airborne division be dropped by parachute behind enemy lines to attack them from the rear. Mitchell thought the airplane's greatest value lay in its ability to strike at centers of population. No longer would civilians be immune from the horrors of war; neither women nor children would be spared. The battle would be carried right to the politicians who were responsible for starting it. Massive bombing raids over Berlin, Mitchell believed, would quickly bring Germany to its knees.

The war ended with Mitchell's plans unrealized and his superiors still skeptical of the airplane's offensive utility. In fact, the plane had done nothing to convince disbelievers that it would ever become a decisive military force. Surely it would be imprudent, when dealing with the nation's security, to place undue faith in Mitchell's unproven bombers or to heed his pleas for an air corps—with its high costs for administration, training, and construction.

For Mitchell and his band of dedicated followers, the battle on behalf of the airplane was just beginning. To dramatize his contention that great surface fleets of battleships would soon be rendered obsolete because of their vulnerability to air attack, Mitchell, with army and navy brass looking on, staged an elaborate experiment off Virginia in 1921. Several captured German warships were towed out to sea and sunk by plane-delivered bombs. Although the test garnered much favorable publicity for the cause, the military hierarchy remained unmoved.

Full of frustration, Mitchell took to publicly upbraiding his superiors for their alleged incompetence and myopia. In 1925 he was court-martialed and convicted of insubordination at one of the most famous military tribunals in history. For his indiscretions, Mitchell was suspended from active service for five years. It took a disaster the size of Pearl Harbor, sixteen years later, to vindicate his views.

Skeptical

Inclined to question or doubt new theories and facts.

Contention

Point or statement that one has argued for.

Jack Dempsey

Boxing Achieves Status as a

In the early twentieth century, prizefighting was still a largely disreputable and, in some places, illegal diversion. The acknowledged "world champions" were less athletes than brawlers. No holds were barred. Rabbit and kidney punches, biting, and gouging were perfectly acceptable techniques. Bouts went on for as many as a hundred rounds, or until one or both of the bare-fisted gladiators collapsed from exhaustion or injury. So the "sport," such as it was, languished in richly deserved notoriety.

By 1920, however, boxing had become a major spectator sport. A British peer, the Marquis of Queensbury, set down rules and regulations, limiting rounds to three minutes and providing for padded leather mittens to cover the knuckles. Prizefighting was rescued from its sordid background; all it needed was a figure to generate excitement and publicity. This came in the form of a six-foot, 185-pound West Virginian named Jack Dempsey who became heavyweight champion in 1919.

Dempsey was one of those champions that the spectators either loved or detested. Part of the negative feeling derived from Dempsey's indictment in 1920 on charges of draft evasion. His manager, the wily Jack "Doc" Kearns, not inclined to have his fighter wind up a target on the French battlefields, had managed to wangle him a draft-deferred job in a shipyard during World War I. Though Dempsey was subsequently cleared of wrongdoing, charges of cowardice plagued him throughout his career.

Perhaps that explains his ferocity in the ring. Dempsey went into every bout with the single-minded aim of battering his opponent into oblivion. That made him a hero in some eyes, and a brutish villain in others. Whatever people thought of him, they turned out in unprecedented numbers to see him either vanquished or victorious. In 1921, his initial title defense against the dapper Frenchman Georges Carpentier drew 91,000 spectators and the first million-dollar gate in boxing history.

On the night of September 14, 1923, at New York's Polo Grounds,

Notorious

Commonly or well-known, especially because of something bad.

Evasion

Escaping; avoiding.

Spectator Sport

Dempsey took on the Argentine Luis Angel Firpo, "The Wild Bull of the Pampas," in what boxing experts have called the Battle of the Century. In round one Firpo was knocked down six times and Dempsey twice. Just before the end of the round Firpo sent Dempsey crashing through the ropes. Dempsey recovered and in round two dropped the Bull to the canvas for the full count.

In 1926 Dempsey lost his crown to Gene Tunney in a ten-round decision in Philadelphia. The next year a rematch was held in Chicago which 145,000 people paid a record sum of $2.6 million to attend. In the seventh round Tunney seemed ready to fall before Dempsey's onslaught. But Jack refused to retreat from his opponent; and, as the referee kept motioning him away, Tunney gained precious seconds to recover before the count officially began. He rose shakily, but won by decision. The "Battle of the Long Count" ended Dempsey's chances of again wearing the crown, but he boxed for another five years, taking on over 175 opponents, and knocking out more than 100 of them.

Babe Ruth

Leisure Time Promotes Sports

Americans in the postwar years suddenly and pleasantly found themselves with new leisure time. In most industries, the six- or seven-day work week was becoming a thing of the past. Annual vacations of a week to a month were commonplace by 1920, even for the lowliest machine operator. With the cuts in working time came greater time in which to spend money. The national mood was one of living life to the fullest, of enjoying the pleasures which had once been available only to the affluent. As promoters found, the sprawling new leisure class was willing to spend, and spend prodigiously, on recreation and entertainment.

Prodigious

Huge; very great; vast.

One place where Americans were spending more of their time and money was the ball park. The twenties saw a spectacular growth in a variety of spectator sports, but baseball's preeminence was widely acknowledged. Attendance at ballgames climbed steadily from 1920 on. Baseball truly became the national pastime, with millions of fans all over the country following the fortunes of their favorite players and teams.

No player captivated the public or did more to revolutionize and popularize the game than George Herman (Babe) Ruth. The product of a Baltimore orphanage, Babe made it to the big leagues on his strong left arm, starting as a pitcher with the Boston Red Sox. His explosive bat, however, made him too valuable to be playing only every third or fourth day, so he was shifted to the outfield. Ruth's stock in trade was the home run: with one mighty stroke of his bat, the ball soared beyond the reach of fielders, beyond fences, beyond the confines of any stadium. The Babe seemed to disdain puny singles or doubles; it was either the glorious satisfaction of a home run or the abysmal failure of a strikeout. Every time he stepped up to the plate the air was filled with excitement and a tense showdown with the pitcher ensued. The "Bambino" emerged the victor as often as not.

Puny

Petty; not important.

With Ruth's remarkable prowess, and his flair for showmanship, it seemed that he was somehow not meant to remain in sedate Boston. In 1920 the Yankees of the American League purchased his contract

90

and Its Heroes

and brought him to New York. The Yankees shared the Polo Grounds in Upper Manhattan with their National League rivals, the Giants. But with Babe socking home runs at an amazing clip, people turned out in numbers beyond the capacity of the Polo Grounds. Thus arose "the House that Ruth Built:" Yankee Stadium, the largest sports arena in the world when it was dedicated in 1923. On opening day, seventy-five thousand fans were present and the Babe rose to the occasion, christening the new ball park and delighting his admirers with a home run.

The year Ruth came to New York he belted fifty-four home runs, a new record. It was the first of a host of records which were to fall before his murderous swing. In his career, which spanned a little more than a decade, he would hit 714 home runs, a mark which stood for nearly fifty years. He hit forty or more homers in each of eleven seasons, reaching sixty in 1927. It began Yankee domination of the American League, which saw them win six pennants and three World Series. Babe became the toast of the town, an authentic hero of his day, even if he neither looked, spoke, nor acted like one. Sportswriter Paul Gallico fondly recalled Ruth as "the big burly man with the ugly face, blob nose, cigar stuck out of the side of his mouth. His speech is coarse, salty, with 'son of a bitch' so frequently, genially and pleasantly used that it loses all of its antisocial qualities, and becomes merely another word that does not particularly disturb." He violated every training rule in sports and was still able to come through with a dramatic home run when the situation called for it. He had a huge appetite for food and liquor and very much enjoyed the company of women. He was more than just a home run hitter, more than just a gifted athlete. All over the league it was an occasion when Babe and his able teammates came to town. The Babe was a symbol of the free-spirited decade, and America loved him for it. Children revered him, and grown men wept when he struck out. For ten years he was baseball; no player before or since has ever dominated the game with the same personal magnetism as Babe Ruth.

Harlem Renaissance

The Flowering of an Urban Community

While European immigrants poured into the country after 1900, a massive internal migration of equal significance was taking place. Streaming into the burgeoning cities of the North were thousands of blacks, abandoning Jim Crow and the economic stagnation of the South. They came in search of jobs and the social equality long withheld from them in the old Confederacy. Many got work—in war plants, in the steel, meat, and automotive industries—but many more found that the dreamed-of city ("streets paved with gold") was tarnished with racism as virulent as that which they had left. During the war years, race riots broke out in Washington, Chicago, St. Louis, and a host of other cities large and small. White America was not at all prepared for the influx of blacks into hitherto white neighborhoods and, with jobs at a premium in the postwar reconversion period, strongly resented the competition of blacks in the marketplace. Since blacks were generally willing to work for wages lower than scale, they were often used as strikebreakers by employers trying to suppress union activity—a fact which made them anathema to organized labor. Indelible immigrants, blacks found asylum behind ghetto walls.

But if the term "ghetto" implies the confinement of a beaten people, then surely it is a misnomer in this case. For the twenties was the age of the new Negro: proud, independent, and determined to fight for what was rightfully his. Blacks, having served their country with distinction during the war, were no longer willing to submit to the indignities of inferiority. In the South they surprised the resurgent Ku Klux Klan by refusing to be brow-beaten. In the urban riots, blacks fended off the attacks of the rampaging white mobs, and even retaliated in kind. Their actions in defense of home and family made it clear that black America had every intention of entering the mainstream.

Hope was what sustained black communities throughout the country. The largest and most varied of them was New York City's Harlem. The story of Harlem's rise to prominence as the capital of black

America contains all the elements then active in growing cities. Harlem was a new, entirely white, middle-class community in 1910, situated on the northern outskirts of Manhattan. It had been built up with large apartment houses, but because the city's mass transit lines did not reach into East Harlem, landlords there found themselves saddled with many vacancies. Almost in desperation they began renting to blacks. Whites in western Harlem began to panic and flee. Property values dropped in the wake of the exodus, permitting poorer blacks to join their growing community.

A true community it was, attracting blacks from Africa and the West Indies as well as from the South. By the twenties Harlem had become a prosperous, law-abiding, culturally rich, and extraordinarily diverse city-within-a-city. It was a monument to black achievement. Its people represented a cross section of American society: students, intellectuals, artists, and businessmen coexisting with peasants, criminals, and exploiters. The better elements gave Harlem its soul and its purpose. With the patronage of white intellectuals and philanthropists, the arts flowered. In verse, Countee Cullen, James Weldon Johnson (a founder of the National Association for the Advancement of Colored People), and Langston Hughes alternately sang the celebration of black life and the despair of its plight in white America. Black musicians took New Orleans jazz and gave it a distinctively urban flavor. Intellectuals such as Alain Locke searched for the blacks' proper role in American society. Marcus Garvey, the Black Moses, preached black pride, black capitalism, and black separatism, and won millions of converts to his Universal Negro Improvement Association. Sociologists W. E. D. DuBois and E. Franklin Frazier won world-wide reputations.

Yet Harlem was an oasis in a desert of white supremacy. Much of its prosperity derived from the patronage of the cafe society who flocked to the Cotton Club or one of the other Harlem cabarets that catered exclusively to whites. Harlem was their playground, but once blacks no longer entertained, they quickly became expendable.

Vacancy

Space, room or apartment for rent.

Plight

Situation or condition, usually bad.

95

Sacco and Vanzetti

Anarchism and Massive Public

To Americans, radicalism always meant violence. They remembered well the Haymarket Riots and the assassination of President McKinley by a self-proclaimed anarchist. Anarchist actions were always dramatic, always defiant, and almost always violent. Anarchy was what tainted radicalism. The great despair of socialists and communists in America was that the public made few distinctions when it came to appraising the left wing: the anarchist element was shunned in the radical community no less than in conventional quarters. That most anarchists also happened to be foreigners (usually Eastern Europeans) tended to make the anarchist an especially repugnant creature. Indeed, the drive to restrict immigration from those areas fed on the fear of radicals and anarchists. The celebrated case of Nicola Sacco and Bartolomeo Vanzetti is illustrative of these concerns and prejudices.

Repugnant

Highly distasteful; objectionable; offensive.

On April 15, 1920, a double murder took place at South Braintree, Massachusetts. Returning from Boston with the $15,000 payroll of the shoe factory for which they worked, Frank Parmenter and Alexander Berardelli (paymaster and guard, respectively), were gunned down and robbed by two men. The thieves made their getaway, but not before several people caught glimpses of them. They were described as having swarthy complexions and probably of Italian origin.

Garnered

Gathered.

The crime garnered little publicity. Three weeks later, two Italian workmen—Sacco and Vanzetti—were arrested and charged with it. In the trial, presided over by Judge Webster Thayer, both men proclaimed their innocence. Both furnished solid alibis and brought a host of character witnesses to testify in their behalf. Neither had any criminal record. The prosecution's evidence connecting them to the crime was largely circumstantial, yet the jury found them guilty and Judge Thayer sentenced them to death.

Two circumstances worked against them. First, they were Italians and aliens. Second, both were avowed anarchists and had taken part in a mass protest against the alleged persecution of radicals by the Department of Justice. Sacco and Vanzetti's political views, more

Concern

than any evidence presented in court, seemed responsible for their conviction.

It is likely that Sacco and Vanzetti would have gone quietly to their deaths had the radical press not picked up the story and distributed it in Europe and South America. All over the world people expressed their anger at what appeared to be a miscarriage of justice (defense lawyers were accumulating new evidence to prove their innocence). A bomb exploded at the home of the American ambassador in Paris; the flag was trampled on in Rome; in Uruguay, American goods were boycotted. Intellectuals like Anatole France, H. G. Wells, and Albert Einstein flocked to their cause, demanding that the case be given a full review. A defense fund, to which radicals and moderates alike subscribed, was raised to defray legal costs.

The two condemned men sat in jail for six years as their case was appealed again and again. Throughout the ordeal, Sacco and Vanzetti won the admiration of even their adversaries for maintaining their composure and dignity under the great stress. Neither seemed capable of committing the kind of heinous crime with which they were charged. Yet in 1927, Thayer stubbornly denied the last appeal, and Sacco and Vanzetti were sent to the electric chair.

Defray

To arrange the payment of.

Rudolf Valentino

A New American Institution:

Whatever their status as an art form, the movies, more than any other consumer product, was intended to make a profit. In the end all that mattered was whether the public would pay at the box office. Like any merchant the filmmaker had to tailor his product according to the marketplace. During the Jazz Age Americans registered a definite preference for sex and scandal on the screen, and the producers complied with large helpings of the lurid.

Indeed, titillation seemed to be Hollywood's chief stock-in-trade in the twenties. Producers found that films about promised sex drew long lines of moviegoers, while straight dramas and patriotic epics played to near-empty houses. Thus, the country received a steady diet of movies with suggestive titles such as *Her Purchase Price*, *A Shocking Night*, and *Up in Mable's Room*. Ad writers outdid themselves with claims of the "truth—bold and sensational."

The sex symbol—the actor who made women swoon or the actress who made men pant—became an American institution in the twen-

The Sex Symbol

ties. They were worshiped, imitated, and lusted for by moviegoers, and their lives were given enormous exposure. To gain maximum publicity for the stars and to encourage patronage of their films, film press agents concocted wild stories about their glamorous exploits and origins. From a tailor's daughter named Theodosia Goodman came the Vamp. Theda Bara—a name which could be respelled Arab Death—was said to be of an Egyptian mother and a French father, and to have grown up learning the secrets of the Orient, which gave her an evil power over men. And from an olive-skinned, twenty-six-year-old immigrant with the formidable name of Rudolf Alfonzo Raffaele Pierre Filbert Guliemi di Valentina d'Antonguolla was created Rudolf Valentino, the epitome of a mysterious sheik.

Valentino was the most popular of them all. Women's hearts were set aflame by his piercing eyes and passionate air, and there were incidents of fainting when the sheik rescued the heroine from the blazing Sahara. During six tumultuous years in Hollywood, in films like *The Four Horsemen of the Apocalypse* and *Blood and Sand*, Valentino set the standards for male sex appeal. Men professed bewilderment at this appeal, but they aped his looks and habits nonetheless. Barbers reported that more men were using pomades to slick down their hair, and tango schools did a booming business.

Off camera, Valentino's life—and even his death—seemed staged entirely for the benefit of the public. The simple act of his growing a beard became a spectacle: women and barbers protested, and there was a great public "debearding" performed by members of the Master Barbers of America. He drove an expensive, fire-engine-red convertible and lived in a hillside marble-floored hacienda known as Falcon's Lair. As quickly as he made a fortune, he spent it. He was so extravagant that by the time of his death in 1926 he was heavily in debt. His funeral drew a crowd which stretched for eleven New York City blocks. Scores of people were injured fighting to catch a glimpse of the fallen sheik in his casket. Valentino was as sensational in death as he was in life.

Scopes Trial

Monkey Business in Tennessee

In the 1920s rural America made a last-ditch attempt to stop the continued spread of dangerous ideas from the cities. It was bad enough that country folk were emigrating in droves to the cities in search of work and excitement. But now urban culture, in the form of movies, radio, and advertising, was invading the hinterlands and was finding a receptive audience. Exposed to the excitement of metropolitan life, farm families were becoming more and more unhappy with the austere, work-filled dawn-to-dusk lives that they led. The old rural belief, which identified hard work with virtue, was being weakened. It its place came a desire to sample and savor all that life had to offer. Upholders of rural values tried in vain to stop the spread of this new infection. Preachers described the terrors of hell and damnation that awaited those who gave into temptation. Nevertheless, their congregations dwindled. Immigration restriction and prohibition were attempts to protect the traditional morality, but they were not enough. Although scapegoats came and went, there was no question as to who was the real enemy of rural culture—the big cities. They were seen as breeding grounds for sin and corruption that spread beyond city limits.

In the face of this disturbing dislocation, the church remained the last bastion of the old morality. A lot of people had stopped attending services, but those who went retained a basic and devout faith that divine guidance was to be found only in the Scriptures. They believed that the real danger to their way of life came from those who challenged the Bible, for to do so was to undermine faith in God's moral rules as well. Science was a chief culprit.

In the cities, science and technology were moving toward the creation of a world where all of man's problems could be solved. In the country, however, people were not so ready to accept man-made solutions and human intervention. In urban schools children were being taught that, contrary to what the Bible said, the lessons of science indicated that man was descended from the apes (this was Charles Darwin's theory of evolution). Rather than have their chil-

Hinterland

Interior or back country; land behind a coastal district.

Bastion

A stronghold; a fortified location.

dren's minds poisoned by what they regarded as nonsense, in the 1920s Fundamentalist Protestants launched an attack on the teaching of the Darwinian theory of evolution. In the Northeast, the anti-evolutionists made no headway. But in the strongholds of Fundamentalism, they were more successful. Under the leadership of William Jennings Bryan, a folk hero in rural America, these religious crusaders won partial victories in Oklahoma, Florida, and North Carolina. Then the Tennessee state legislature passed a law making it illegal "for any teacher in any of the . . . public schools of the state to teach any theory that denies the story of the divine creation of man as taught in the Bible and to teach instead that man has descended from a lower order of animals." This law led to one of the most famous trials in history.

The man about whom the controversy would rage was John T. Scopes, a young biology teacher at Dayton High School. Without any idea that his actions would have such far-reaching consequences, he had lectured on Darwin's ideas of evolution and natural selection. He was arrested and the case was brought before a grand jury.

The trial began in July 1925. The attornies for the defense were Clarence Darrow, the most famous lawyer of the day, and Arthur G. Hays, an eminent civil liberties attorney. Bryan himself assisted the prosecution, whose case depended only on proving what Scopes freely admitted—that he had broken the law. The defense sought to show that the law was unjust and foolish; that the Bible should be taken as an allegory, not as the literal truth; and that a belief in evolution was not contrary to the Christian faith. Scopes was found guilty, as expected, and fined, but the verdict was an anticlimax after the drama that had unfolded when Bryan was called to the stand by the defense, as an "expert" on the Bible. Under Darrow's devastating questioning, the farmer's champion was revealed to be narrow and foolish. The antievolutionists had had their day, but in the end they had to give way—cosmopolitanism and science were too much for them.

Controversy

An argument; a long dispute.

Eminent

Distinguished; above all others.

Bootlegging

A New Form of Crime Results

It has been said that the American people are most resourceful in emergencies. Adversity has always seemed to bring out the same enterprising spirit of pioneer self-sufficiency and intensity of purpose that enabled earlier generations to subdue the wilderness. Americans have never willingly submitted to infringements of their personal liberties; indomitably, they have bounded back and emerged victorious in the end.

In light of these much-celebrated national qualities, the attempt to deprive the nation of alcohol after the First World War was doomed to be an exercise in futility. It was almost as if those supporting prohibition had set out with the secret intention of putting American traditional qualities to the test—to see whether the American gift for ingenious evasion of unpopular laws was still alive. And at the end of the great experiment, weary enforcement officers knew that the American spirit of resistance was alive and well. At no time in American history has drinking been a more fashionable pastime than during the era of prohibition. For those who wanted it, liquor was as widely available as it had ever been, although from different sources. Bootlegging—the illegal manufacture and transport of alcohol—had become what was probably the nation's most profitable industry.

Like any other big business, bootlegging had its conglomerates and independents. Anyone with the inclination, a few hundred dollars, and a spare room could become an instant entrepreneur. The economics were enticing. Depending on fluctuations in supply and demand, home-distilled bootleg (as opposed to the "imported" variety) sold for between five and fifteen dollars a quart. A commercial still, with the capacity to turn out four hundred quarts a day, could be purchased for as little as five hundred dollars. For the less venturesome, portable models were available for about five dollars. Overhead was nonexistent, and there were no taxes to pay. With only local (and bribable) police and a woefully undermanned force of largely inept federal agents to worry about, confiscation or arrest

Indomitable

Unyielding, that which cannot be conquered.

Entrepreneur

A successful businessman.

from Prohibition

was improbable. In fact, the prohibition commissioner himself admitted that his men were able to intercept only five percent of the illicit traffic in alcohol.

The big boys had most of the clout. Before the ink had dried on the prohibition legislation, organized crime, never an institution to pass up lucrative opportunities, was retooling to supply the black market in bootleg. By 1925 the mob controlled an estimated ninety percent of the illegal liquor and beer business in the United States, amounting to over a quarter of a billion dollars a year. Smugglers were commissioned to bring in Canadian and Scotch whiskies, which were watered down and sold to the well-heeled. Industrial alcohol, made almost palatable by being combined with flavorings, was the poor man's drink, but no less profitable for the mob. Bribes were considered regular business expenses; in several cities the entire municipal government was on the take. Americans could well deplore organized crime's role in bootlegging, but it is noteworthy that they did so while lifting their glasses.

Flappers

Women's Liberation and the Postwar

The First World War wrought substantial changes in American life. Not the least of these was a revolution in morals and manners that, more than anything else, gave the postwar decade its distinctive flavor. Victorian ideas about behavior came under assault; the nation's youth seemed embarked on a campaign to popularize behavior that flew in the face of standards long established by the society. The older generation, full of concern, searched feverishly for an explanation as to why the country, fresh from its overseas crusade, was falling to such depths of depravity. Some accused the Bolsheviks (members of the Communist party) of spreading atheism and anarchy among the susceptible American youth. But the worriers need not have looked for foreign agents to take the blame for the rapid social change in America. It was inevitable that the cream of American manhood would return from Europe anxious to relive the pleasures of the cafés and Parisian nightlife. And American women, having shed their petticoats and rolled up their sleeves to work in the wartime economy at jobs that were traditionally held by men, emerged from the war with a heightened sense of self-esteem and a desire for greater personal and sexual freedom.

The Twenties might be called the decade of the feminist. Women began afresh to examine their role in society and to assert their right to make their own decisions on matters that concerned them. And during these years women made great advances in breaking the political, moral, and economic shackles that bound them in a male-dominated world. With women's suffrage a long-awaited reality in 1920, inroads began to be made in erasing the more archaic aspects of the laws governing male-female relationships. Women were accorded more legal recognition as full partners in marriage. The divorce rate spiraled, as increasing numbers of women sought freedom from marriages which they felt had narrowed and somehow diminished them. Puritanical and discriminatory moral strictures went by the boards in many cases. Women were no longer arrested for smoking a cigarette on Fifth Avenue, as they had been in 1904; nor

Depravity

A practice or an act showing corrupted morals.

Shackle

Anything that prevents freedom.

Morality

was it uncommon for women to mingle freely with men at cocktail parties. The national mood was one of self-indulgent abandon, and women meant to share in the mood and in the prosperity of the Twenties.

The most outspoken and unreserved women appeared to delight in flaunting their new freedoms before a scandalized America. The liberated woman's conversation, often sprinkled with four-letter words, was as likely to run to sexual matters as to recipes. The most militant feminists, like Charlotte Perkins Gilman, spoke out forcefully against all forms of male oppression. Many women shed their corsets for less restrictive, more suggestive attire—short clinging dresses, silk stockings rolled below the knees, high heels, and swimwear that exposed more flesh than it covered. Thus was born the flapper, the symbol of the age. She might be seen strolling on big city streets—hair bobbed, cheeks rouged, catching admiring and disdainful glances. Or she could be found in a dance hall, fox-trotting to the wails of the saxophone; and at wild campus parties she drank bootleg liquor from hip flasks. She frequented speakeasies, football games, and risqué movies.

But the flapper, like bathtub gin, was only a part of the larger revolution in morality that was taking place. Despite the sharp disapproval of provincial America practically no aspect of American life remained untouched by the changed social climate. The arts especially showed its influence. In films, "sex symbols" like Rudolph Valentino and Theda Bara titillated audiences; in literature Ernest Hemingway and F. Scott Fitzgerald wrote with a new bluntness; and jazz, the music of the backstreets of New Orleans, was the rage. The flapper—the independent, not so demure young American woman—seemed to sum up all that was happening in the Twenties. But there remained of course another America out in the countryside and in the small cities that resisted the new wave and stuck firmly, in fact aggressively to its traditional ideas of morality, religion, and women's place in society.

Flaunt
Show off.

Disdainful
Scornful and proud.

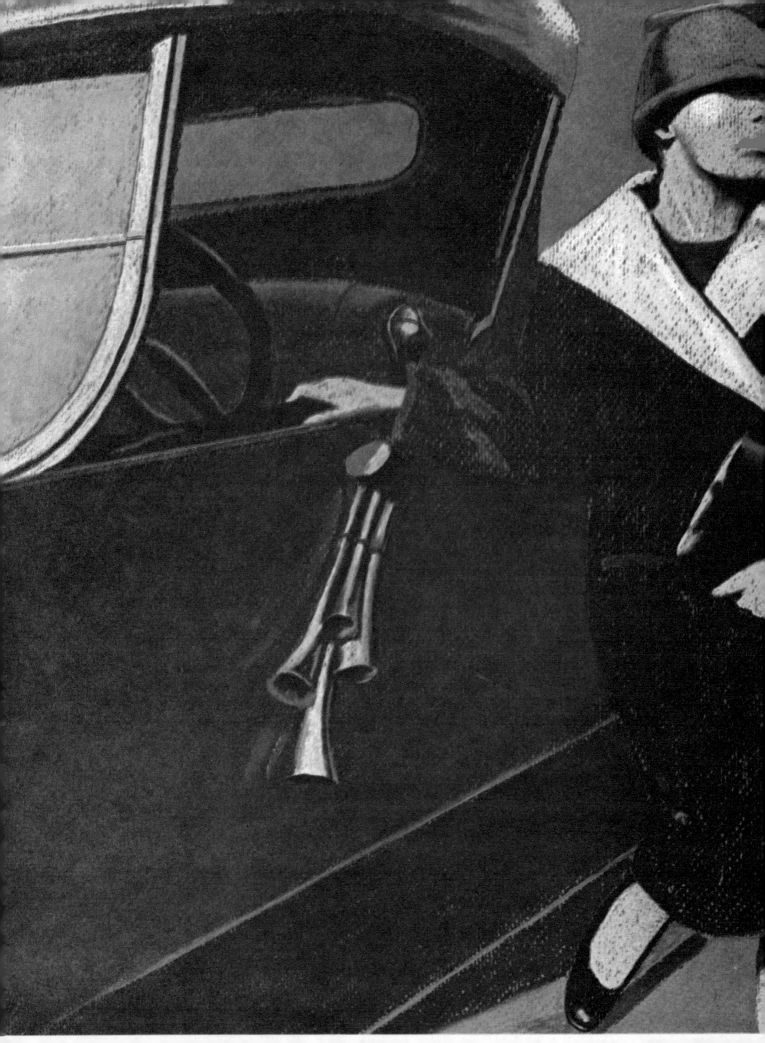

Red Scare

Effects of the Threat of Revolution

Despotism

The exercise of absolute authority. An autocratic or absolute form of government.

In November 1917, after centuries of Czarist despotism, the Russian people revolted. Led by Nikolai Lenin and Leon Trotsky, they proclaimed a Union of Soviet Socialist Republics, the first state founded on the communist principles of Karl Marx and Friedrich Engels. To Westerners, the overthrow of capitalism in Russia was serious enough in itself, but what was most worrisome was that the Bolshevik leaders predicted, and took active steps to bring about, a worldwide socialist revolution. In less turbulent times, the possibility of such a catastrophe would have been minimal. The year 1918, however, found Europe in shambles and ripe for revolution. Treasuries had been depleted by four years of total war, leaving even the victors unable to rebuild their devastated countries. Workers took to the streets throughout Europe in a wave of violent general strikes, inspired by the Russian example. In Hungary and in parts of Germany, the legitimate governments were toppled by communist uprisings. Poland and Italy were on the brink of falling.

Deplete

To reduce or decrease the amount of.

The United States was undergoing difficulties as well. Although it emerged from the war relatively unscathed, the economy was undergoing the painful process of reconversion to peacetime operation. Orders for war goods were abruptly canceled, leaving workers idle and machines silent. Returning servicemen added to the unemployment problem. During the war, wages and prices had been regulated by law. But the controls were lifted soon after the signing of the armistice. As a result, the prices of consumer goods skyrocketed, with salaries unable to keep pace. The cost of living rose 99 percent between 1914 and 1919, and the 1919 dollar was worth less than half of the value of the 1913 dollar.

Intrusion

Coming unwanted and unasked.

The government's solution was to leave matters alone. Businessmen were in no mood to put up with controls which they believed were an unreasonable intrusion into a free market economy. Determined to make up ground lost during the war, the unions assumed a more militant policy in dealing with employers. When their demands were rebuffed, labor leaders were not reluctant to call walkouts.

on the United States

That labor won but a handful of those struggles is in large part due to the impact of the revolutionary activity in Europe. Americans realized that the Atlantic Ocean afforded them no protection from the spread of revolutionary ideas. Bolsheviks bragged openly about the amounts of money being expended to prepare the United States for eventual take-over. Membership in the two American communist parties jumped to seventy thousand by 1919, and radicals like John Reed were confident that revolution in this country was imminent.

Part of the left-wing optimism derived from the false belief that American labor had concluded that all attempts to work within the system were futile, and therefore would take up the cause of revolution. The American middle class seemed to agree that this change had taken place already. Their suspicions were confirmed when a general strike, not unlike those which preceded revolution in Europe, paralyzed the city of Seattle, where Mayor Ole Hanson falsely blamed the strike on the communists. The Seattle strike made a strong impression on a frightened public, which tended from then on to see all labor disturbances as the start of revolutionary activity.

Labor was further discredited when, in the spring of 1919, a series of random bombings against public officials took place. Although it is known that they were the work of a few fanatical anarchists, the country demanded swift retaliation against all individuals with "un-American" tendencies. President Wilson's attorney general, A. Mitchell Palmer, was given the task of rooting all radicals. In what was probably the single greatest violation of civil liberties in American history, thousands of suspected extremists were rounded up and herded to detention centers. Palmer's raids turned up almost nothing in the way of arms, and most of the people seized were completely innocent of wrongdoing or of radical connections. These facts and the realization that the communist danger had been grossly exaggerated brought the country back to its senses and ended the Great Red Scare. But it had been an unnerving example of public hysteria and official disregard for legality and the rights of individuals.

Farmers' Co-ops

A Way to Get Out of the Agricultural

The twenties can be seen as a study in economic contrasts. On the one hand, there was the prosperity of business and industry, which trickled down to the individual worker in the form of an improved standard of living. The worker now had more free time to spend with his family and more money to spend on cars and other luxury items. The cities swelled with consumers eager to spend their money. Optimism abounded; city life in the twenties seemed to be the fulfillment of the American dream.

Conditions on the farm, however, were quite different. Farmers were experiencing the worst depression in a generation. It came suddenly, after twenty years of unparalleled prosperity. In no comparable period had the farmer been as well off as he was in the years 1900-20. Shortages all over the world, made even more acute by World War I, created a situation where the demand for foodstuffs far outstripped supply, driving agricultural prices to dizzying heights. The bushel of wheat that brought the farmer sixty-two cents in 1900 was worth more than four times as much nineteen years later. Farmers responded to the boom in the usual way: they cultivated more intensively while prices remained high.

In 1920 the bottom dropped out of the commodities market, mainly because world demand dropped sharply once the war was ended. The American farmer, who found himself holding enormous surpluses, was caught in the vise of the law of supply and demand. By June 1921 the farmer was receiving an average of forty to fifty percent less per bushel than in the previous year. Many farmers, having lost everything, deserted their lands and moved to the city where there were better opportunities. For the first time in American history farm acreage decreased.

Apart from the unusual intensity of this decline, there was actually nothing new in the farmer's predicament. Boom and bust were his lot. Lately, however, he had become more insistent in his demands that the government help ease his plight. Farm elements converged on their congressmen, seeking legislation that in essence would do one

Depression

of two things: allow the government to subsidize farmers when prices dropped below a certain level; or provide foreign markets where farmers could sell their surplus. Both measures would mean considerable expense to the government and the urban-dominated, economy-minded Congress was reluctant to go along. When the farm bloc finally managed to push the McNary-Haugen bill through Congress, President Calvin Coolidge vetoed it.

The president, along with Secretary of Commerce Herbert Hoover, had another idea. Coolidge thought that rather than rely on the government to pull them out of a situation they had created through reckless overproduction, farmers should help themselves by establishing cooperatives and discouraging overproduction.

The cooperative seemed like a good idea. By acting together to keep a particular crop off the market when prices were too low for them to make a reasonable profit, farmers would conceivably have a hand in controlling the cycles of supply and demand. In practice, however, the cooperative had one crucial weakness: unanimity of all the producers of a commodity was necessary for it to be effective. Some farmers were unwilling to give up their independence and the cooperative failed as a solution to the problem.

Surplus

Extra quantity; amount left over.

Muscle Shoals

One Consequence of the "Hands-off"

After weathering twenty years of trust-busting under the progressive Presidents Roosevelt, Taft, and Wilson, business found the Republican administrations of the twenties ready and willing to do its bidding. In that decade, government listened attentively and, more often than not, acted obligingly when the business community spoke. What it said can be condensed into one phrase: "hands off." Business insisted that government withdraw from any participation in the free enterprise system. It should relinquish any control of business and dissolve constricting regulatory agencies so business could function freely and profitably. Of course, any assistance government could provide in the way of protective tariffs or antilabor measures would be most welcome.

Relinquish

Let go; give up.

The administration's position in the Muscle Shoals controversy illustrates the extent of its willingness to abide by business's wishes. It also illustrates the courage and determination of Senator George Norris of Nebraska, who almost singlehandedly thwarted the business-government alliance on the question of federal ownership of utilities.

Thwart

Keep from doing something; oppose and defeat.

Muscle Shoals, the Alabama town on the Tennessee River, was the site of two plants built by the government in 1916 to supply nitrates (the key ingredient in gunpowder and fertilizer) for the war effort. The plants extracted nitrogen from the air, using a procedure which required huge amounts of electricity, so the government had also begun work on a 100-foot dam and a power plant. But before the project reached completion, the war concluded, thus ending the urgent need for nitrates. Though the U.S. had already spent nearly $150 million at Muscle Shoals, Congress cut off federal funds for construction of the Wilson Dam (as it was later called), bringing work to a halt. At the same time, the War Department solicited bids from private concerns to take over the government complex for private exploitation.

Henry Ford saw that Muscle Shoals had great potential for profit. With the quantity of hydroelectric power it would produce, he could

Policy

supply all the needs of the Tennessee Valley farmers and still have enough left for his own automobile factories. Land prices shot up in anticipation; no one doubted that Congress would give Ford what he wanted.

Senator Norris, chairman of the committee to which the Ford offer was submitted, had become convinced that the resources of the valley should be developed by the government for the best interests of all the people. Norris saw that Muscle Shoals could furnish inexpensive power to the factories, homes, and farms of the area. Much to the disgust of Ford and the valley farmers—who were ready to cash in on the Ford scheme— Norris was able to have it shelved. He finally won congressional support for completion of the Wilson Dam—only to see it pocket-vetoed by President Coolidge. Later, a similar bill was vetoed by Herbert Hoover. Norris had saved Muscle Shoals from the clutches of private industry, but not until 1933, when Franklin D. Roosevelt created the Tennessee Valley Authority (TVA) were some of his dreams put into action.

Lindbergh

The First Transatlantic Air Journey

On the gloomy Friday morning of May 20, 1927, a lanky, intensely handsome, twenty-five-year-old aviator climbed into the tiny cockpit of a silver-winged monoplane poised on Long Island's Roosevelt Field. The aviator was Charles A. Lindbergh; despite his youth he had years of flying experience as a member of the Army Air Service Reserve and as a stunt and airmail pilot. The plane was *The Spirit of St. Louis*, elegant in its simplicity and designed with a single purpose—to carry one man across the Atlantic to France.

Since the Wright brothers had flown their first "motorized glider" over the dunes at Kitty Hawk in 1903, aviation technology had made enormous strides, transforming the airplane from a novelty to a feared weapon of war and a fast, reliable means of long-distance transport. But no union of man and plane, it seemed, could hurdle the barrier of the Atlantic. Enticed by a standing prize of twenty-five thousand dollars to the first successful transatlantic pilot, many tried and failed, often losing their lives in the attempt. The problem lay in the enormous quantity of gasoline needed for a transatlantic flight. No one had been able to build a plane combining huge fuel capacity with airworthiness. Laden with as much as three tons of gas and crews of up to four men, most planes were unable to even make it off the runway.

The Spirit of St. Louis embodied Lindbergh's solution to the problem. It was a model of stripped-down efficiency; everything not absolutely essential to the plane's operation was discarded. Windshield and parachute were both conspicuously absent, as was a heater to provide warmth at freezing altitudes. The pilot was to sit strapped in an uncomfortable, but lightweight wicker chair, encircled by tanks of gasoline. Lindbergh knew that just one spark of lightning would turn the cockpit into an inferno from which there was no escape. He would fly alone, without navigator or radio, with only a compass and a few maps to guide him. For his own use he carried only a canteen of water, seven sandwiches, and a small survival kit containing a flashlight, some matches, and a rubber raft.

That Friday morning a small crowd of sleepy-eyed spectators gathered to witness Lindbergh's departure. The crowd held its breath as *The Spirit of St. Louis* started its takeoff. The plane rose reluctantly, settled down with a sickening skid, and rose again, this time for good, on its way to an uncertain destiny.

The trip was not without its tribulations. In his excitement Lindbergh had not slept at all the night before his departure. Now, at ten thousand feet, fatigue began to weigh heavily upon him. Not daring to leave the controls for even a minute of rest, Lindbergh had to hold his eyes open with his hands and once slapped himself in the face with full force to stay awake. Bad weather was a constant worry. About two hundred miles east of Newfoundland he ran into an ice storm which nearly sent him to a grisly death in the frigid North Atlantic. But Lindbergh adeptly maneuvered the craft out of danger and continued on to his destination. And he made it, some thirty-three and a half hours after takeoff. Landing at Paris's Le Bourget Airport at about 10 P.M., *The Spirit of St. Louis* was mobbed by more than one hundred thousand wildly cheering Frenchmen, who manhandled Lindbergh and tore at the plane for souvenirs. Wherever he went, Lindbergh was surrounded by fans. Strangers clustered around him, begging for an autograph or a memento.

The moment he touched down, Lindbergh became an instant hero. The American public was intoxicated by Lindbergh's great adventure; his daring, courage, and fierce independence of spirit seemed to symbolize all that was great about the United States. He was a shining inspiration for a people that had seemingly lost its sense of purpose in the tawdry Twenties. The country was swept up in celebration of "Lucky Lindy." Streets, towns, and babies were named after him. Four million New Yorkers turned out to welcome him home with a ticker-tape parade the likes of which no one had ever seen before. And in other cities, the receptions for this pioneer who had conquered a great frontier were no less frenzied. The nation rushed to make Lindbergh and his achievement its own.

Destiny

What becomes of a thing or a person in the end.

Intoxicated

The act of becoming greatly excited.

Jimmy Walker

A Showman in Charge of

New York City in the twenties was the nation's pleasure capital. When rural people criticized the cities for their decadence, they had the Gotham example in mind. This was the festive era when New York became the Big Apple, where dissipation was fashionable and no one cared to deny it. The city had the biggest and most exclusive speakeasies for the fashionable elite of show business and just plain millionaires. It had gambling dens and Harlem during its renaissance. And it had Jimmy Walker as mayor.

James J. Walker was his name, but no one called him that. Walker was always Jimmy, the witty, corrupt mayor-about-town; the actor, comedian, civic pleasure-seeker who symbolized the holiday spirit and the vivacity and splendor of New York during the Jazz Age. The mood of the day demanded entertainment more than government, and Walker dutifully played his part. For the majority of New Yorkers who because of poverty or limited means were unable to partake of the high life, Walker was everyman's playboy. He regaled in luxury provided by an indulgent

118

New York City

public. He was befriended by gangsters; he junketed with millionaires at Palm Beach and Paris. New Yorkers apparently were so amused by his antics that they elected him to a second term. But when hard times came, Walker's only prescription for recovery was that the movie theaters should show nothing but cheerful films. His government was incapable of providing leadership when it was needed. By 1929, with countless citizens out of work and Hoovervilles rising in Riverside Park, Walker was a disturbing anachronism, and his reign was brought to an inglorious end.

Walker had had a taste of show business in his early years as a song writer on Tin Pan Alley. He seemed to have had better training there than in the state legislature, which he entered as no less than a reformer. He adopted a faintly progressive platform in his run for the mayoralty in 1925. During the campaign he brought tears to the eyes of his listeners when he declared: "Out of a heart filled with precious memories of my home and parents, if I am elected mayor, New York City will be the best, cleanest, most wholesome, and orderly place in America." He was elected, of course, but he was aware that civic virtue was fast becoming obsolete. He loved New York, but he wanted to get away from it and its problems as much as he could; fully one-quarter of his time in office was spent at fashionable resorts in the United States and overseas. His press conferences were akin to vaudeville performances. His work day typically commenced at noon, and there were times when he never made it to City Hall. Even his critics admitted that Walker had a keen mind and an ability to grasp complex problems quickly, but he used it more often for personal benefit than for governing.

By 1929 Walker could no longer dismiss corruption with a joke. Governor Franklin D. Roosevelt ordered a judicial investigation of city mismanagement. The investigating committee recommended that Walker be impeached. Before that could happen, Walker resigned from office and sailed for Europe in disgrace. He was one of the many casualties of the Great Depression.

Al Smith

The Voice of Urban America

Ascend

Rise; go up.

The career of Alfred E. Smith is a remarkable success story. The son of poor Irish immigrants, he became the Democratic candidate for president in 1928. That someone like Smith, a school dropout at fifteen, a city boy, and a Roman Catholic, could ascend to the top of the political heap speaks volumes about the changes that were overtaking America in the 1920s.

Smith's ladder to fame was Tammany Hall, New York's Democratic political machine. For eight years he served the organization faithfully as a process server. He was rewarded for his loyalty with a seat in the state assembly. Smith soon compiled a brilliant legislative record and a reputation that promised greater achievements ahead. In 1918 he won the governorship of New York handily and began what would be an eight-year stay in Albany. Under the leadership of Governor Smith, with strong bipartisan backing, New York enacted some of the nation's most far-reaching social legislation. Smith displayed courage in standing up to the forces of fear and repression during the postwar Red Scare and he was most skillful in dealing with the perennial Republican majority in the legislature.

Smith's prominent position among urban politicians put him into the national spotlight in 1924. He had a certain rough charm about him. He wore his origins like a badge—the touch of an Irish brogue, the brown derby hat, and the cigar butt jammed between pursed lips became a symbol not only of Smith, but also of all big-city politicians.

Stature

Accomplishment; development.

The Democratic party, looking for a man of stature, seized upon Smith as a likely presidential candidate. But at the 1924 convention, a stalemate developed between Smith, representing the urban, anti-prohibition wing of the party, and William G. McAdoo, who, though also a New Yorker, was favored by the rural South and West. The delegates chose a compromise candidate—John W. Davis, a distinguished but colorless lawyer. Even before the ticket had gone down to a crushing defeat, the Smith forces were readying themselves for 1928. Smith was clearly the darling of the 1928 convention and won its nomination with great acclaim. His Republican opponent was Her-

bert Hoover of Iowa, a man of great ability and wide experience.

It was an uphill battle for Smith. The Republicans could boast of eight years of unparalleled prosperity under the Harding and Coolidge administrations. Hoover ran on a platform that promised more of the same. Smith, on the other hand, sought to appease everyone—farmers, workers, and industrialists—and in trying to do so pleased no one. Since there was little difference between the candidates' stands on real issues, the campaign's focus was on religion, prohibition, and personality.

Appease
Satisfy.

From the start Smith's Catholicism was the target of an ugly campaign of slurs and innuendos. His foes conjured up wild images of the pope dominating the White House. And the organized forces of evangelical Protestantism opposed his decidedly wet stance. When the votes were counted, Hoover emerged with a substantial majority. The vote against Smith was partially a vote against the cities and all the alien values they seemed to represent.

Herbert Hoover

The Logical Republican Successor

When Calvin Coolidge declared laconically in 1927, "I do not choose to run," an alliance of Republicans and businessmen set out to find a mutually acceptable candidate for the next year's presidential race. They did not have to look far. The natural choice was Herbert Hoover of Iowa. No man had a better reputation for honest, selfless, capable public service. His career was legendary. An orphan at the age of eight, through stern self-discipline he had risen to eminence in both business and government. As a mining engineer and later a partner in his own firm, he traveled and made friends all over the world. He was a millionaire at forty. During World War I he served so brilliantly as head of the U.S. Food Administration and the Belgian relief effort that, according to one observer, he was "the only man who emerged from the ordeal of Paris with an enhanced reputation." When he returned home, only his claim that he was uninterested stopped the Hoover-for-president talk from carrying him to the White House. Appointed secretary of commerce by President Warren Harding, he remained in that office for seven years. Business never had a more loyal or able friend. Hoover embodied all the virtues of the efficient, public-spirited, cooperative capitalist. In 1928 the Republican Convention nominated him on the first ballot.

Eight years of Republican prosperity—along with Hoover's background, personal integrity, and ability on the stump—made him an assured winner for president against the controversial Alfred E. Smith. Hoover and his running mate, Charles Curtis of Kansas, won a landslide victory. Everyone believed that their election meant a continuation of economic growth and good times.

Hoover was in office only seven months when in October 1929 the stock market crashed and the Great Depression set in. At first it was thought that the Depression would be short-lived and that recovery would come in a few months. By the beginning of 1930, however, such views were shattered, as stock and commodity prices plunged even lower. Entire fortunes were wiped out, banks closed, and businesses failed, putting millions of people out of work.

to Coolidge

Hoover had not yet heeded the obvious warning signals of collapse. He had remained confident that the business structure was sound and that it only needed slight stimulation from government to return to normal. He relied on the weight of his office to convince businessmen to maintain employment and wages and to increase investments, but he did little to compel their compliance. And when misery spread over the land, he took few steps to relieve the people's plight. Hoover disliked the idea of government relief (he feared it would stifle initiative and create a costly bureaucracy); so he left the problem of feeding and sheltering the vast numbers of unemployed and their families to private charities and local authorities, who were unable to handle the burden. All the country's bitterness was directed at the ineffectual man in the White House. Shanty towns were dubbed Hoovervilles, and he became the object of scorn and derision. By 1932 the country's patience had run out, and Franklin D. Roosevelt won the presidency in a landslide victory.

H. L. Mencken

The Captain at the Helm of

Intellect

Power of understanding; knowing.

The 1920s were America's most culturally productive years since the 1850s. The literary output of the twenties was largely a response by the intellectual class to what it saw as a spiritual decay in society. Emerson, Whitman, and Thoreau, among the "lost generation" of young thinkers in the fifties, were revolted by what they viewed as the preoccupation with material things and corruption of their times. Whereas the transcendentalists of the fifties were confident in man's capacities to bring about progress, the intellectuals of the twenties were not nearly so optimistic. They had been largely disappointed by the accomplishments of the Progressive Movement and now concluded that the system was hopelessly corrupt and the people lost. Human nature, they decided, had brought progress, but if progress meant the ugliness and greed of America's business civilization, they would have none of it. Cynical and disillusioned, many took up residence in Europe, where they found a climate more conducive to creativity.

Expatriate

To withdraw from one's native country.

The spokesman in America for the expatriate intellectuals was Henry L. Mencken. Mencken, a Baltimorean trained as a journalist, had always been an outspoken and witty critic of the American scene. As literary critic and co-editor of the magazine *Smart Set*, Mencken took on anything that smacked of hypocrisy or self-righteousness. When that magazine folded, Mencken, along with George Jean Nathan, began in 1923 to publish the *American Mercury*, a periodical addressed to the intellectual left wing. Every month it carried, in addition to reviews of books and the theater, original works by members of the new wave of angry young writers, many of whom had been laboring in what they considered ill-deserved obscurity. Theodore Dreiser, Frank Norris, Sherwood Anderson, and Willa Cather, among others, found a sponsor in Mencken and a voice in the *Mercury*.

By far the most popular feature of the *Mercury* was Mencken's commentary on American life. No institution was safe from his gibes. He took special delight in tormenting the low-brow majority, who to

Literary America

him were the "booboisie." The mass of Americans, he wrote, were "the most timorous, sniveling, poltroonish, ignominious mob of serfs and goosesteppers ever gathered under one flag in Christendom since the end of the Middle Ages." The barbarism of the mob had transformed democracy, Mencken believed, into a ludicrous hoax. He was "against all theologians, professors, editorial-writers, right-thinkers, and reformers." He proposed legalizing assassination of public officials, abolishing the public school system, and dumping the Statue of Liberty into the ocean. He held nothing sacred. Mencken assaulted the manners and pride of the "boobocracy" of small-town executives everywhere, informing them, on one occasion, that all Anglo-Saxons were cowards. Country folk came in for their share of invective when Mencken covered the Scopes trial. He referred to the Tennessee farmers as "gaping primates" and "the anthropoid rabble"; they were "yokels, hillbillies, and peasants."

The *Mercury* took college campuses by storm. For more than ten years Mencken titillated his readers with his outrageous banter. With the *Mercury's* circulation reaching nearly 80,000 by 1928, Walter Lippmann could rightfully call Mencken "the most powerful influence on this whole generation of educated people."

Timorous

Timid; easily frightened.